Compressibility *of* Ultra-Soft Soil

Compressibility of Ultra-Soft Soil

Myint Win Bo
B.Sc, DUC, M.Sc, Ph.D, FGS, FICE,
C.Geol, C.Eng, C.Sci, C.Env, P.Geo, Eur Geol, Eur Eng

DST Consulting Engineers Inc., Canada

NEW JERSEY · LONDON · SINGAPORE · BEIJING · SHANGHAI · HONG KONG · TAIPEI · CHENNAI

Published by

World Scientific Publishing Co. Pte. Ltd.
5 Toh Tuck Link, Singapore 596224
USA office: 27 Warren Street, Suite 401-402, Hackensack, NJ 07601
UK office: 57 Shelton Street, Covent Garden, London WC2H 9HE

British Library Cataloguing-in-Publication Data
A catalogue record for this book is available from the British Library.

COMPRESSIBILITY OF ULTRA-SOFT SOIL

Copyright © 2008 by World Scientific Publishing Co. Pte. Ltd.

All rights reserved. This book, or parts thereof, may not be reproduced in any form or by any means, electronic or mechanical, including photocopying, recording or any information storage and retrieval system now known or to be invented, without written permission from the Publisher.

For photocopying of material in this volume, please pay a copying fee through the Copyright Clearance Center, Inc., 222 Rosewood Drive, Danvers, MA 01923, USA. In this case permission to photocopy is not required from the publisher.

ISBN-13 978-981-277-188-9
ISBN-10 981-277-188-3

Typeset by Stallion Press
Email: enquiries@stallionpress.com

Printed in Singapore.

Dedicated to my late parents U Bo Byu and Daw Thein

PREFACE

Due to the scarcity of land at coastal regions around the world, land reclamation is commonly carried out for the future expansion of various infrastructure facilities. Soft soil is present at most of the coastal regions of Southeast Asia and thus land reclamation on this highly compressible soil foundation often requires the use of soil improvement works to eliminate significant future settlements from occurring. The combination of prefabricated vertical drains with preloading is one of the most widely used ground improvement methods in land reclamation projects.

Due to the rapid expansion of major coastal cities like Singapore, the requirement for land reclamation is critical in such city states due to the scarcity of land. Often, areas that have been used as dumping location for dredged spoils, mine tailing activities or waste ponds require land reclamation to cater for the rapid expansion of the population as well as for major industries. Material in such type of ponds are extremely soft.

The deformation behaviour of ultra-soft soil due to additional load is different to that of the normally or overconsolidated soft soils. The prediction of magnitude and time rate of deformation by use of conventional soil mechanics theory is not applicable for such ultra-soft soils. The deformation behaviour of such ultra-soft soil deposits was previously not well understood nor the behaviour of such deposits when subject to land reclamation fills or as a result of ground improvement. The formation of such soil has to be studied from the sedimentation process right to the completion of self-weight consolidation by means of field tests and specialized laboratory tests.

This book is a result of extensive years of research and practical experience gained by the author who was actively involved in the land reclamation and ground improvement works at Changi East in the Republic of Singapore. The project involved works where such ultra-soft soil was encountered in the project and had to be treated with prefabricated vertical drains to enable the reclaimed land to be used for major airport facilities.

The author spent many years working on the design and implementation at the project and embarked on an illustrious research career researching in topics such as land reclamation, soft soil engineering, in-situ testing, laboratory testing, field instrumentation, ground improvement and sand compactions works. Extensive specialized laboratory tests were carried out by the author and this was validated by soil models and field instrumentation results. A theory for settlement prediction for such ultra-soft soils was also developed by the author during the course of this research as was a finite-difference model for ultra-soft soils. This book should prove useful to the practicing engineers, project managers and researchers who are involved in land reclamation and ground improvement projects as well as students and contractors alike. The works by the author are land mark works in a new field of study which has never before been studied and thus this book now provides extensive inroads on the field of ultra-soft soil.

Dr A. Vijiaratnam
Former Chairman SPECS Consultants Pte Ltd, Singapore
Former Pro-Chancellor Nanyang Technological University, Singapore

CONTENTS

Preface vii
List of Tables xv
List of Figures xix
List of Symbols xxxiii
Acknowledgements xxxix
About the Author xli

1. Introduction   1
   1.1 Deformation of Soils and Their Compressibility   1
   1.2 Related Previous Research   2
   1.3 Outline of the Book   4

2. Sedimentation and Consolidation   6
   2.1 Sedimentation   6
   2.2 Sedimentation to Consolidation 11
   2.3 Self-Weight Consolidation 14
   2.4 Theory of Self-Weight Consolidation 15
   2.5 Large Strain Consolidation 20
   2.6 Compression and Consolidation of Ultra-Soft Soil under Additional Load 20
   2.7 Prediction of Magnitude and Rate of Settlement in Large Strain Consolidation 20
       2.7.1 Magnitude of Settlement 20
       2.7.2 Time Rate of Settlement 22

3. Models and Analogy   24
   3.1 Viscoplastic Model for Ultra-Soft Soil 24

	3.2		Modified Spring Analogy for Ultra-Soft Slurry-Like Soil .	25
	3.3		Magnitude and Rate of Compression	26
		3.3.1	Magnitude of Compression	26
		3.3.2	Combined Basic Equation for Viscous Stage and Soil Stage	27
		3.3.3	Time Rate of Compression	29
		3.3.4	Slurry to Soil	30
		3.3.5	Coordinate System and Governing Equation . . .	30
			3.3.5.1 Coordinate System	30
			3.3.5.2 Governing Equation	31
		3.3.6	Numerical Approach	32
4.	Characterization of Physical Properties and Mineralogy of the Soil			34
	4.1		Introduction	34
	4.2		Physical Properties	35
	4.3		Mineralogy of Clay	36
	4.4		Chemistry of Pore Water	38
	4.5		Preparation of Ultra-Soft Sample	39
5.	Compression Tests with Large Scale Consolidometer			43
	5.1		Method of Testing	43
	5.2		Description of Apparatus and Sample Preparation	44
	5.3		Two-Step High Pressure Loading Tests	46
		5.3.1	Deformation Behavior During First Step Loading	46
		5.3.2	Pore Pressure and Settlement	47
		5.3.3	Comparison of Degree of Consolidation	53
		5.3.4	Deformation Behavior During Second Step Loading	54
		5.3.5	Post-Mortem Investigation on the Compressed Sample	55
			5.3.5.1 Moisture Content and Bulk Density . .	61
			5.3.5.2 Preconsolidation Pressure	62
			5.3.5.3 Undrained Shear Strength	62
	5.4		Incremental Step Loading Test	63
		5.4.1	First Step Loading	64
		5.4.2	Second Stage Loading	69
		5.4.3	Third Step Loading	74
		5.4.4	Post-Mortem Investigation on Compressed Sample	78
	5.5		Summary	82

6.	Compression Test on Slurry with Small-Scale Consolidometer		84
	6.1	Description of the Apparatus	84
	6.2	Method of Testing	85
	6.3	Discussion on Test Results	85
		6.3.1 Settlement Response	85
		6.3.1.1 Effect of Initial Moisture Content	85
		6.3.1.2 Effect of Magnitude of Applied Pressure	100
		6.3.2 Pore Pressure Response	106
		6.3.2.1 Effect of Initial Moisture Content	107
		6.3.2.2 Effect of Magnitude of Applied Pressure	109
		6.3.3 Comparison of Pore Pressure from Transducers Located at Various Locations	113
	6.4	Compressibility and Consolidation Parameters in the Soil Stage	113
	6.5	Discussion on Void Ratio at Transition Point	123
	6.6	Verification of Non-Homogeneity During Slurry Compression	125
	6.7	Laboratory Tests on Compressed Samples	126
	6.8	Summary	126
7.	Compression Tests on Ultra-Soft Soil with Hydraulic Consolidation Cell		129
	7.1	One-Step High Pressure Loading Tests with Hydraulic Consolidation Cell	129
		7.1.1 Description of the Apparatus	129
		7.1.2 Sample Preparation and Method of Testing	130
		7.1.3 Discussion of Results	130
		7.1.3.1 Pore Pressure Response	130
		7.1.3.2 Settlement Response	136
		7.1.3.3 Rate of Settlement and Hydraulic Conductivity	136
		7.1.4 Relationship Between Void Ratio and Applied Additional Load	142
		7.1.4.1 Effective Stress Gain and Compression Index (C_{c1}^*) in Viscous Stage	144
		7.1.4.2 Transition Void Ratio	150
	7.2	Step Loading Compression Tests on Ultra-Soft Soil with Hydraulic Consolidation Cell	151

		7.2.1	Method of Tests	151
		7.2.2	Discussion of Results	152
			7.2.2.1 Settlement Characteristics	152
			7.2.2.2 Pore Pressure Characteristics	155
		7.2.3	Correlation of Compression Index with Physical Parameters	156
		7.2.4	Validation of Basic Equation for Magnitude of Settlement	164
		7.2.5	Coefficient of Consolidation for Large Strain	169
	7.3	Summary		176
8.	Continuous Loading Tests on Ultra-Soft Soil			178
	8.1	Constant Rate of Loading Test on Ultra-soft Soil		178
		8.1.1	Apparatus Used	178
		8.1.2	Discussion of the Results	180
			8.1.2.1 Settlement and Pore Pressure Behavior	180
			8.1.2.2 Effective Stress Gain at the Base and Average Effective Stress	183
			8.1.2.3 Void Ratio Change versus Average Effective Stress Gain	188
			8.1.2.4 Compression Indices	188
			8.1.2.5 Variation of e-log σ' Curve with α Values	192
			8.1.2.6 Hydraulic Conductivity and Large Strain Coefficient of Consolidation	192
	8.2	Constant Rate of Strain Test on Ultra-Soft Soil		195
		8.2.1	The Apparatus	196
		8.2.2	Sample Preparation and Test Method	196
		8.2.3	Selection of Strain Rate	197
		8.2.4	Discussion on the Test Results	198
			8.2.4.1 Behavior of Pore Pressure and Applied Load	198
			8.2.4.2 Excess Pore Pressure Ratio	200
			8.2.4.3 Effective Stress Gain	202
			8.2.4.4 Compression Indices	203
			8.2.4.5 Coefficient of Consolidation and Hydraulic Conductivity	203
	8.3	Summary		206

9. Verification of Proposed Formulae and Models with
   Laboratory Measurements 208
   9.1 Prediction of Magnitude and Time Rate of Settlement .. 209
       9.1.1 Magnitude of Settlement 209
             9.1.1.1 Gibson et al. (1981) 209
       9.1.2 Time Rate of Settlement 212
             9.1.2.1 Gibson et al. (1981) 212
             9.1.2.2 Cargill (1982) 215
             9.1.2.3 Lee and Sills (1981) 216
             9.1.2.4 Proposed model 222
   9.2 Comparison of Various Large Strain Existing Models
       with Proposed Time Factor Curve 227
   9.3 Summary 235

10. Case Study 237
    10.1 Description of the Silt Pond 237
         10.1.1 Soil Condition in the Silt Pond 238
         10.1.2 Characterization of the Silt Pond Slurry Prior to
                Reclamation 238
    10.2 Reclamation of the Silt Pond 242
         10.2.1 First Phase Sand Spreading 242
         10.2.2 Site Investigation During and After First Phase
                of Sand Spreading 243
         10.2.3 Failure During Sand Spreading 247
         10.2.4 Remedial Measures 247
         10.2.5 Second Phase of Sand Spreading 248
         10.2.6 Intermediate Mini-Cone Penetration Tests 249
         10.2.7 Characterization of Soil after Sand Spreading 250
         10.2.8 Lowering of Ground Water Level 252
         10.2.9 Installation of Vertical Drains and Surcharges ... 253
    10.3 Pilot Embankment and Other Test Areas 253
    10.4 Deformation Behavior of the Silt Pond Slurry 254
    10.5 Interim Verification of Improvement of Slurry 258
         10.5.1 Verification from Analysis of Settlement Data 258
         10.5.2 Verification from Analysis of Piezometer
                Monitoring Data 258
         10.5.3 Verification Using *In Situ* Tests 259

10.6　Verification of the Proposed Large Strain
　　　　　Deformation Model . . . .　261
　　　　　10.6.1　Determination of Compression Indices . . . . . .　264
　　10.7　Verification Using Data from Area with the Largest
　　　　　Deformation in the Main Works . . . .　268
　　10.8　Verification of Large Strain Deformation using Data
　　　　　from No-Drain Area. . . . .　271
　　10.9　Summary . . . .　273

11. Summary　　　　　　　　　　　　　　　　　　　　　　　　274

　　11.1　Sedimentation and Consolidation . . . .　274
　　11.2　Experimental Studies . . . .　275
　　11.3　Validation of Proposed Finite Difference Model . . . . .　276
　　11.4　Case Study . . . .　277

References　　　　　　　　　　　　　　　　　　　　　　　　　279

LIST OF TABLES

Table 4.1.	Relative consistency (Cr) and liquidity index (LI) of tested soils.	36
Table 4.2.	Chemistry of pore water from tested samples.	41
Table 5.1.	Basic physical properties of tested soils.	45
Table 5.2.	Summary of measurements in two-step high pressure loading.	47
Table 5.3.	Comparison of time at various degrees of consolidation based on the settlement and pore pressure measurements.	54
Table 5.4.	Comparison of strain from subdivided layers at the end of tests.	60
Table 5.5.	Effective stress gain of soil element at different locations.	63
Table 5.6.	Comparison of c_u measured in the laboratory and calculated applying study of Ladd *et al.* (1977) for two-step high pressure loading test.	64
Table 5.7.	Summary of measurements in step incremental loading.	69
Table 6.1.	Convergent point based on settlement rate.	91
Table 6.2.	Time and void ratio at convergence determined from merging of void ratio curves.	93
Table 6.3.	Time and void ratio at convergence determined from hydraulic conductivity of ultra-soft soils.	95
Table 6.4.	Hydraulic conductivity change indices of various moisture content soils at various stages.	99
Table 6.5.	Comparison of transition point determined from settlement and time when pore pressure dissipation commences.	113

List of Tables

Table 6.6.	Comparison of void ratio at transition points determined from commencement of pore pressure dissipation.	117
Table 6.7.	Initial, transition, and final void ratio of various initial moisture content soils under various loadings.	118
Table 6.8.	Comparison of compressibility parameters at soil state.	124
Table 7.1.	Initial condition of sample and applied pressure.	132
Table 7.2.	Comparison of applied pressure, responded pore pressure and duration of delayed in pore pressure dissipation.	132
Table 7.3.	Comparison of transition void ratio determined by various methods.	148
Table 7.4.	Comparison of compression indices (assuming interception at e_{10}).	148
Table 7.5.	Details of tests.	151
Table 7.6.	Details of loadings.	152
Table 7.7.	Comparison of magnitude of settlement and change in void ratio in each step.	154
Table 7.8.	Comparison of compression parameters.	155
Table 7.9.	Comparison of predicted and measured settlements.	166
Table 8.1.	Details of tests.	179
Table 8.2.	Peak pore pressure and void ratio.	185
Table 8.3.	Variation in α with b/r (after Smith and Wahls, 1969).	188
Table 8.4.	Comparison of measured and calculated compression parameters ($\alpha = 0.667$).	191
Table 8.5.	Comparison of measured and calculated compression parameters from selected rate of loading ($1\,\text{kPa}/800\,\text{s}$) applying selected α value of 0.75.	194
Table 8.6.	Suggested strain rate for high moisture content soil.	198
Table 10.1.	Soil properties of silt pond clay prior to reclamation and after sand spreading.	242
Table 10.2.	Summary of compressibility parameters for ultra-soft upper and lower soil layer at silt pond area.	264
Table 10.3.	Summary of predicted settlement under various stages at silt pond pilot area.	265
Table 10.4.	Summary of applied C_F values in the prediction of time rate of settlement in silt pond pilot area.	268

Table 10.5.	Summary of compressibility parameters for ultra-soft upper and lower soil layer at silt pond large settlement area.	269
Table 10.6.	Summary of settlement under various stages at silt pond large settlement area.	270
Table 10.7.	Summary of applied C_F values used in the prediction of time rates of settlement in silt pond large settlement area.	271
Table 10.8.	Summary of compression parameters for ultra-soft soil in the containment bund, no-drain area.	272
Table 10.9.	Summary of applied C_F values and predicted settlement at the no-drain area.	272

LIST OF FIGURES

Fig. 2.1.	Regimes of sedimentation (after Fitch, 1979).	8
Fig. 2.2.	Sedimentation zone in continuous thickening (after Fitch, 1979).	9
Fig. 2.3.	Sedimentation which takes into account the consolidation (after Bustos, 1987).	10
Fig. 2.4.	Idealized transition point between suspension and soil (after Hight et al., 1987).	13
Fig. 2.5.	Profile of effective stress increment with time from sedimentation test on high initial solid concentration slurry (solid concentration = 40%) (after Li and Williams, 1995).	16
Fig. 2.6.	Isochrones of void ratio: (a) after Lee (1979) or Eq. (2.8); (b) expected from experiments (Been and Sills, 1981).	17
Fig. 2.7.	Boundary conditions and definition of the modified problem (after Been and Sills, 1981).	18
Fig. 2.8.	Isochrones of void ratio when $z_0 = 1.5\,z_1$ (after Been and Sills, 1981).	18
Fig. 2.9.	Isochrones of excess pore water pressure when $z_0 = 1.5\,z_1$ (after Been and Sills, 1981).	19
Fig. 2.10.	Comparison of theoretical model with surface settlements and real densities for various duration (after Been and Sills, 1981).	19
Fig. 2.11.	Void ratio-effective stress relationship for very soft clay (after Tan et al., 1988).	21
Fig. 3.1.	Summary of existing soil models.	25
Fig. 3.2.	Viscoplastic model for ultra-soft soil.	25
Fig. 3.3.	Magnitude and rate of compression.	26

List of Figures

Fig. 3.4.	Comparison of e-log σ' curves for natural soil and ultra-soft slurry-like soil.	28
Fig. 3.5.	Lagrangian and convective coordinates: (a) initial configuration at $t = 0$, (b) configuration at time "t" (after Gibson et al., 1967).	32
Fig. 4.1.	Typical grain size distribution of tested sample.	35
Fig. 4.2.	Physical characteristics of study soil on characteristics chart.	35
Fig. 4.3.	X-ray diffraction results from 2-m depth ultra-soft soil.	37
Fig. 4.4.	X-ray diffraction results from 5-m depth ultra-soft soil.	38
Fig. 4.5.	X-ray diffraction results from 8-m depth ultra-soft soil.	39
Fig. 4.6.	Magnification of soil fabric under electron microscope.	40
Fig. 4.7.	Comparison of moisture content with and without salt corrections.	41
Fig. 5.1.	Large diameter consolidometer equipped with pore pressure transducers and total pressure cell.	46
Fig. 5.2.	Pore pressure measurement during the first step high pressure loading: (a) mini-PWP 3; (b) mini-PWP 4; (c) PWP 1; (d) PWP 2; (e) mini-PWP 1; (f) mini-PWP 2. (1 and 2 refer to semi-log scale and arithmetic scale respectively.)	48
Fig. 5.3.	Vertical settlement measurement during the first step high pressure loading.	52
Fig. 5.4.	Total pressure measurements during the first step high pressure loading.	53
Fig. 5.5.	Pore pressure measurement during the second step high pressure loading: (a) mini-PWP 1; (b) mini-PWP 2; (c) mini-PWP 3; (d) mini-PWP 4. (1 and 2 refer to semi-log scale and arithmetic scale, respectively.)	56
Fig. 5.6.	Vertical settlement measurement during the second step high pressure loading.	59
Fig. 5.7.	Shape of PVD and location of mini-PWP after the completion of the compression test.	59
Fig. 5.8.	Closed up view of the compressed sample.	60
Fig. 5.9.	Cutting of samples from the compressed block sample.	60
Fig. 5.10.	Grain size distribution of samples after compression.	61
Fig. 5.11.	Measured moisture contents and void ratios after compression. ($53/1.4 =$ Moisture content/Void ratio.).	61

Fig. 5.12.	Measured preconsolidation pressure in kPa after compression.	62
Fig. 5.13.	Measured undrained shear strength in kPa after compression.	63
Fig. 5.14.	Pore pressure measurement during the first step incremental loading: (a) PWP 1; (b) PWP 2; (c) mini-PWP 1; (d) mini-PWP 2; (e) mini-PWP 3; (f) mini-PWP 4. (1 and 2 refer to semi-log scale and arithmetic scale, respectively.)	65
Fig. 5.15.	Total pressure measurements at the top and bottom pressure cells during the first step incremental loading.	69
Fig. 5.16.	Pore pressure measurement during the second step incremental loading: (a) PWP 1; (b) PWP 2; (c) mini-PWP 1; (d) mini-PWP 2; (e) mini-PWP 3; (f) mini-PWP 4. (1 and 2 refer to semi-log scale and arithmetic scale, respectively.)	70
Fig. 5.17.	Total pressure measurements at the top and bottom pressure cells during the second step incremental loading.	74
Fig. 5.18.	Pore pressure measurement during the third step loading: (a) mini-PWP 1; (b) PWP 2; (c) mini-PWP 2; (d) mini-PWP 3; (e) mini-PWP 4. (1 and 2 refer to semi-log scale and arithmetic scale, respectively.)	75
Fig. 5.19.	Total pressure measurements at the top and bottom pressure cells during the third step incremental loading.	78
Fig. 5.20.	Vertical settlement measurement during the first, second, and third step high pressure loading.	79
Fig. 5.21.	Measured moisture content and void ratio after compression test.	79
Fig. 5.22.	Measured preconsolidation pressure in kPa after compression test.	80
Fig. 5.23.	Measured undrained shear strength in kPa by lab vane after compression.	81
Fig. 5.24.	Correlation between moisture content and undrained shear strength.	81
Fig. 6.1.	Small-scale consolidometer equipped with pore pressure transducers.	86

List of Figures

Fig. 6.2.	Settlement (strain) versus time for various moisture contents under the same loading of 127 kPa. (a) arithmetic scale, (b) log scale.	87
Fig. 6.3.	Settlement (strain) versus time for various moisture contents under the same loading of 175 kPa. (a) arithmetic scale, (b) log scale.	88
Fig. 6.4.	Settlement (strain) versus time for various moisture contents under the same loading of 223 kPa. (a) arithmetic scale, (b) log scale.	89
Fig. 6.5.	Settlement rate versus time for various moisture contents under the same loading. (a) 127 kPa, (b) 175 kPa, (c) 223 kPa.	90
Fig. 6.6.	Strain rate versus time for various moisture contents under the same loading. (a) 127 kPa, (b) 175 kPa, (c) 223 kPa.	91
Fig. 6.7.	Void ratio versus time for various moisture contents under the same magnitude of load. (a) 127 kPa, (b) 175 kPa, (c) 223 kPa.	92
Fig. 6.8.	Darcy hydraulic conductivity versus time for various moisture contents under the same magnitude of loading. (a) 127 kPa, (b) 175 kPa, (c) 223 kPa.	94
Fig. 6.9.	Void ratio versus Darcy hydraulic conductivity for various moisture contents under the same loading. (a) 127 kPa, (b) 175 kPa, (c) 223 kPa.	96
Fig. 6.10.	Void ratio versus Darcy hydraulic conductivity for samples of the same moisture content soil but under various loadings. (a) $W = 130\%$, (b) $W = 150\%$, (c) $W = 170\%$, (d) $W = 190\%$.	97
Fig. 6.11.	Hydraulic conductivity change index versus initial void ratio.	98
Fig. 6.12.	Settlement (strain) versus time for samples of the same moisture content but under various loadings. (a) arithmetic scale, (b) log scale.	100
Fig. 6.13.	Settlement (strain) versus time for samples of the same moisture content but under various loadings. (a) arithmetic scale, (b) log scale.	101
Fig. 6.14.	Settlement (strain) versus time for samples of the same moisture content but under various loadings. (a) arithmetic scale, (b) log scale.	102

Fig. 6.15.	Settlement (strain) versus time for samples of the same moisture content but under various loadings. (a) arithmetic scale, (b) log scale.	103
Fig. 6.16.	Void ratio versus applied pressure from various compression tests.	103
Fig. 6.17.	Rate of settlement versus time for samples of the same moisture content but under various loadings. (a) $W = 130\%$, (b) $W = 150\%$, (c) $W = 170\%$, (d) $W = 190\%$.	104
Fig. 6.18.	Strain rate versus time for samples of the same moisture content but under various loadings. (a) $W = 130\%$, (b) $W = 150\%$, (c) $W = 170\%$, (d) $W = 190\%$.	105
Fig. 6.19.	Void ratio versus time for sample of the same moisture content but under various loadings. (a) $W = 130\%$, (b) $W = 150\%$, (c) $W = 170\%$, (d) $W = 190\%$.	107
Fig. 6.20.	Darcy hydraulic conductivity versus time for samples of the same moisture content but under various loadings. (a) $W = 130\%$, (b) $W = 150\%$, (c) $W = 170\%$, (d) $W = 190\%$.	109
Fig. 6.21.	Pore pressure versus time for samples with the various moisture contents but under the same loading. (a) 127 kPa, (b) 175 kPa, (c) 223 kPa. *Note: 1 and 2 refer to semi-log scale and arithmetic scale, respectively.	111
Fig. 6.22.	Pore pressure versus void ratio for sample of various moisture contents under the same loading. (a) 127 kPa, (b) 175 kPa, (c) 223 kPa.	114
Fig. 6.23.	Pore pressure versus time for the same moisture content soil but under various loadings. (a) $W = 130\%$, (b) $W = 150\%$, (c) $W = 170\%$, (d) $W = 190\%$. *Note: 1 and 2 refer to semi-log scale, and arithmetic scale, respectively.	115
Fig. 6.24.	Pore pressure versus void ratio for the same moisture content soil but under various loadings. (a) $W = 130\%$, (b) $W = 150\%$, (c) $W = 170\%$, (d) $W = 190\%$.	118

Fig. 6.25.	Pore pressure versus time measured from pore pressure transducers at various locations. (a) $W = 130\%$, $\sigma = 175\,\text{kPa}$, (b) $W = 170\%$, $\sigma = 175\,\text{kPa}$, (c) $W = 170\%$, $\sigma = 223\,\text{kPa}$, (d) $W = 190\%$, $\sigma = 127\,\text{kPa}$, (e) $W = 190\%$, $\sigma = 223\,\text{kPa}$. *Note: 1 and 2 refer to semi-log scale and arithmetic scale, respectively.	120
Fig. 7.1.	Arrangement of hydraulic Rowe cell test.	131
Fig. 7.2.	Pore pressure versus time for samples with the same moisture content but under various loadings $W =$ (a) 130%, (b) 150%, (c) 168%, (d) 190%. (1 and 2 refer to semi-log scale and arithmetic scale, respectively.)	133
Fig. 7.3.	Comparison of settlements measured from LVDT and calculated from volume change.	136
Fig. 7.4.	Variations of settlement versus time for various moisture contents but under the same loading, Applied pressure = (a) $62\,\text{kPa}$, (b) $118\,\text{kPa}$, (c) $230\,\text{kPa}$.	137
Fig. 7.5.	Variations of settlement versus time for the same moisture content but under various loadings $W =$ (a) 130%, (b) 150%, (c) 168%, (d) 190%.	138
Fig. 7.6.	Rate of settlement versus time for samples of various moisture contents but under the same magnitude of loading.	139
Fig. 7.7.	Rate of settlement versus time for samples of the same moisture content but under different applied loads.	140
Fig. 7.8.	Hydraulic conductivity versus time for samples with various moisture contents but under the same loading.	140
Fig. 7.9.	Hydraulic conductivity versus time for samples of the same moisture content but under various loadings.	141
Fig. 7.10.	Void ratio versus hydraulic conductivity for samples of various moisture contents under the same loading.	141
Fig. 7.11.	Void ratio versus hydraulic conductivity for the same moisture content under various loadings.	141
Fig. 7.12.	Comparison of e-$\log k$ relationship between two type of tests under similar loadings. (a) $W = 130\%$, (b) $W = 150\%$, (c) $W = 170\%$, (d) $W = 190\%$	143

Fig. 7.13.	Void ratio versus effective stress for various moisture contents under the same loading. Applied pressure = (a) 62 kPa, (b) 118 kPa, (c) 230 kPa.	145
Fig. 7.14.	Void ratio versus effective stress for the same moisture content soil under various loadings. W = (a) 130%, (b) 150%, (c) 168%, (d) 190%.	146
Fig. 7.15.	C_{c1}^* determined assuming intercept at e_{10}^*.	147
Fig. 7.16.	Initial void ratio versus compression indices in first log cycle (C_{c1}^*).	149
Fig. 7.17.	C_{c1}^* determined assuming intercept of first slope at transition void ratio.	149
Fig. 7.18.	C_{c1}^* determined from intercept at transition void ratio versus pressure.	150
Fig. 7.19.	Diagram showing arrangement of drainage in compression tests.	152
Fig. 7.20.	Variations of vertical strain and settlement with time. W = (a) 150%, (b) 168%.	153
Fig. 7.21.	Variations of void ratio with pressure for samples with various drainage conditions. W = (a) 150%, (b) 168%.	156
Fig. 7.22.	Void ratio versus pressure from two different moisture contents. (a) Inward drainage, (b) outward drainage, (c) inward and outward drainage.	157
Fig. 7.23.	Pore pressure versus time from various loading steps (W = 150%). (a) Inward drainage, (b) outward drainage, (c) inward and outward drainage. (1 and 2 refer to semi-log scale and arithmetic scale, respectively.)	158
Fig. 7.24.	Pore pressure versus time from various loading steps (W = 168%). (a) Inward drainage, (b) outward drainage, (c) inward and outward drainage. (1 and 2 refer to semi-log scale and arithmetic scale, respectively.)	160
Fig. 7.25.	Liquidity versus vertical effective stress for various types of clay (after Carrier and Beckman, 1984) together with tested clay from the present study. . . .	163
Fig. 7.26.	Relationship between e_{10}^* versus e_L.	164
Fig. 7.27.	Comparison of predicted and measured void ratio with pressure (W = 150%). (a) Inward drainage, (b) outward drainage, (c) inward and outward drainage.	167

Fig. 7.28.	Comparison of predicted and measured void ratio with pressure ($W = 168\%$). (a) Inward drainage, (b) outward drainage, (c) inward and outward drainage.	168
Fig. 7.29.	Comparison of predicted and measured settlements.	169
Fig. 7.30.	Relationship between void ratio and hydraulic conductivity.	170
Fig. 7.31.	Degree of settlement (U_s) versus time factor (T_F) for various loadings ($W = 150\%$). (a) Inward drainage, (b) outward drainage, (c) inward and outward drainage.	171
Fig. 7.32.	Degree of settlement (U_s) versus time factor (T_F) for various loadings (W = 168%). (a) Inward drainage, (b) outward drainage.	172
Fig. 7.33.	Comparison of degree of settlement (U_s) versus time factor (T_F) for different moisture contents (inward drainage).	172
Fig. 7.34.	Comparison of degree of consolidation (U_s) versus time factor (T_F) for various drainage conditions ($W = 168\%$).	173
Fig. 7.35.	Comparison of predicted and measured settlements with time, $W = 150\%$ (inward drainage).	173
Fig. 7.36.	Comparison of predicted and measured settlements with time ($W = 150\%$) outward drainage. Loading steps = (a) 12 kPa, (b) 50 kPa, (c) 100 kPa, (d) 200 kPa, (e) 400 kPa.	174
Fig. 7.37.	Comparison of predicted and measured settlements with time, $W = 150\%$ (inward and outward drainage, loading step = 200 kPa).	175
Fig. 7.38.	Comparison of predicted and measured settlements with time ($W = 168\%$, outward drainage). Loading steps = (a) 10 kPa, (b) 20 kPa, (c) 40 kPa, (d) 80 kPa, (e) 160 kPa.	175
Fig. 7.39.	Comparison of predicted and measured settlements with time ($W = 168\%$, inward and outward drainage). Loading steps = (a) 200 kPa, (b) 400 kPa.	176
Fig. 8.1.	Photographic features of CRL and CRS tests.	179
Fig. 8.2.	Comparison of settlement versus time from various rate of loadings. (a) $W = 150\%$, (b) $W = 170\%$.	180

Fig. 8.3.	Comparison of settlement versus time between two different moisture content soils under same rate of loading. Rate of loading = (a) 1 kPa/150 s, (b) 1 kPa/500 s, (c) 1 kPa/800 s.	181
Fig. 8.4.	Comparison of pore pressure versus time from various rate of loadings. (a) $W = 150\%$, (b) $W = 170\%$. . . .	182
Fig. 8.5.	Comparison of pore pressure versus time between two different moisture content soils under same rate of loading. Rate of loading = (a) 1 kPa/300 s, (b) 1 kPa/500 s, (c) 1 kPa/800 s.	184
Fig. 8.6.	Ratio of average pore pressure to pore pressure at the base versus time factor.	187
Fig. 8.7.	Comparison of e-log σ' curves from various loading rate CRL tests with step loading hydraulic cell tests ($W = 150\%$). (a) Inward drainage, (b) outward drainage, (c) inward and outward drainage.	189
Fig. 8.8.	Comparison of e-log σ' curves from various loading rate CRL tests with step loading hydraulic cell tests ($W = 170\%$). (a) Inward drainage, (b) outward drainage, (c) inward and outward drainage.	190
Fig. 8.9.	Comparison of e-log σ' curves resulted from various α values in 1 kPa/800 s, rate of loading tests with step loading hydraulic cell test. (a) Inward drainage, (b) outward drainage, (c) inward and outward drainage.	193
Fig. 8.10.	Void ratio versus hydraulic conductivity.	195
Fig. 8.11.	C_F versus void ratio.	196
Fig. 8.12.	Comparison of settlement versus time between settlement measured from LVDT, volume change, and applied strain rate.	199
Fig. 8.13.	Comparison of measured load versus time in various strain rate. (a) $W = 130\%$, (b) $W = 150\%$.	199
Fig. 8.14.	Comparison of load and pore pressure versus time at strain rate = 0.01%/min. (a) $W = 130\%$, (b) $W = 150\%$.	200
Fig. 8.15.	Comparison of generated excess pore pressure ratio from various strain rate. (a) $W = 130\%$, (b) $W = 150\%$.	201
Fig. 8.16.	Comparison of e-log σ'_v curves from CRS tests with various strain rate and conventional 24 h test curves. (a) $W = 130\%$, (b) $W = 150\%$.	202

List of Figures

Fig. 8.17. Comparison of hydraulic conductivity measured from various strain rate. (a) $W = 130\%$, (b) $W = 150\%$. . . . 205

Fig. 8.18. Comparison of void ratio versus coefficient of consolidation (C_F) from various strain rate (a) $W = 130\%$, (b) $W = 150\%$. 206

Fig. 9.1. Void ratio–effective stress relationship based on Gibson et al. (1967) if λ is assumed constant: (a) $W = 150\%$; (b) $W = 170\%$. 210

Fig. 9.2. Void ratio–effective stress relationship based on Gibson et al. (1967) when λ is assumed constant. (a) Canaveral Harbour Material; (b) Craney Island Material. 211

Fig. 9.3. Variation of λ with effective stress (after Gibson et al., 1967). 211

Fig. 9.4. Predicted versus measured void ratio–effective stress relationship after applying varied λ values: (a) $W = 150\%$; (b) $W = 170\%$. 212

Fig. 9.5. Variation of λ values with effective stress used in the prediction of magnitude of void ratio change applying Gibson et al. (1967) (a) $W = 150\%$; (b) $W = 170\%$. . 213

Fig. 9.6. Time factor curves proposed by Gibson et al. (1981). 215

Fig. 9.7. Variation of N with effective stress ($W = 150\%$). . . . 216

Fig. 9.8. Measured and predicted time settlement curves: Loading step (a) 12 kPa (b) 25 kPa (c) 50 kPa (d) 100 kPa (e) 200 kPa (f) 400 kPa. 217

Fig. 9.9. Measured and predicted time settlement curves: Loading step (a) 10 kPa (b) 20 kPa (c) 40 kPa (d) 80 kPa (e) 160 kPa (f) 320 kPa (g) 640 kPa. 219

Fig. 9.10. Measured and predicted time settlement curves: Loading step (a) 12 kPa (b) 25 kPa (c) 50 kPa (d) 100 kPa (e) 200 kPa (f) 400 kPa. 223

Fig. 9.11. Measured and predicted time settlement curves: Loading step (a) 10 kPa (b) 20 kPa (c) 40 kPa (d) 80 kPa (e) 160 kPa (f) 320 kPa (g) 640 kPa. 225

Fig. 9.12. Measured and predicted time settlement curves applying present study: Loading step (a) 12 kPa (b) 25 kPa (c) 50 kPa (d) 100 kPa (e) 200 kPa (f) 400 kPa. 228

Fig. 9.13.	Measured and predicted time settlement curves: Loading step (a) 10 kPa (b) 20 kPa (c) 40 kPa (d) 80 kPa (e) 160 kPa (f) 320 kPa (g) 640 kPa.	230
Fig. 9.14.	Comparison of time factor curves from the present study and from the study of Lee and Sills (1981) (a) single drainage (b) double drainage.	232
Fig. 9.15.	Comparison of time factor curves from the present study and from Cargill (1984) (a) double drainage (b) single drainage.	233
Fig. 9.16.	Time factor curves, double drainage condition compared with time factor curves from measured data: (a) Cargill, 1984 (b) Lee and Sills, 1981 (c) proposed model.	234
Fig. 10.1.	Location of silt pond.	238
Fig. 10.2.	Twist sampler.	239
Fig. 10.3.	Locations of density logging, boreholes, and field vane.	240
Fig. 10.4.	Contour map showing depth to density $1.3\,g/cm^3$.	240
Fig. 10.5.	Contour map showing depth to density $1.5\,g/cm^3$.	241
Fig. 10.6.	Density profile along Section A–A'.	241
Fig. 10.7.	Grain size distribution curve of silt pond materials.	242
Fig. 10.8.	Silt pond material on classification chart.	243
Fig. 10.9.	Geotechnical parameters versus depth prior to sand spreading: (a) bulk density; (b) liquidity index; (c) water content and Atterberg limit; (d) initial void ratio; (e) compression index; (f) coefficient of consolidation; (g) overburden and preconsolidation pressure; (h) overconsolidation ratio; (i) shear strength from field vane test; (j) sensitivity from field vane test.	244
Fig. 10.10.	Sand spreader used in silt pond reclamation.	245
Fig. 10.11.	(a) Silt pond prior to reclamation; (b) Sand spreading in silt pond in progress.	245
Fig. 10.12.	Hydrographic survey profile after each phase of spreading.	246
Fig. 10.13.	Comparison of density profile after first phase of sand spreading.	247

Fig. 10.14.	Geotechnical parameters versus depth after first phase of sand spreading: (a) bulk density; (b) liquidity index; (c) water content and Atterberg limits; (d) void ratio after sand spreading; (e) compression index; (f) coefficient of consolidation; (g) overburden and preconsolidation pressure; (h) overconsolidation ratio; (i) field vane shear strength; and (j) sensitivity from vane shear test.	248
Fig. 10.15.	Layout of geofabric placement in the center of silt pond.	249
Fig. 10.16.	(a) Geofabric laying over failure area. (b) Geofabric laying in progress.	250
Fig. 10.17.	Mini-seacalf.	251
Fig. 10.18.	Typical cone resistance profile after sand spreading.	251
Fig. 10.19.	Cross-sectional profile showing build up of sand after sand spreading (interpreted from mini-CPT).	252
Fig. 10.20.	Thickness of built up sand after sand spreading interpreted from mini-CPT data.	253
Fig. 10.21.	Comparison of geotechnical parameters before and after sand spreading: (a) bulk density; (b) water content; (c) Atterberg limits; (d) liquidity index; (e) initial void ratio; (f) preconsolidation pressure.	254
Fig. 10.22.	Instrument layout and installed elevations at silt pond pilot area.	255
Fig. 10.23.	Elevation of instruments installed at an area with largest settlement.	255
Fig. 10.24.	Elevation of instruments installed at the no-drain area.	256
Fig. 10.25.	Piezometers with protected guard shell.	257
Fig. 10.26.	Detail of specially protected pneumatic piezometer with extended protection pipe.	257
Fig. 10.27.	Excess pore pressure versus time at silt pond pilot area (Pneumatic Piezometers).	259
Fig. 10.28.	Excess pore pressure versus time at silt pond pilot area (Vibratory Wire Piezometer).	259
Fig. 10.29.	Settlement versus time at silt pond pilot area.	260
Fig. 10.30.	Reduction of load caused by submergence of land compared with excess pore pressure from vibrating wire piezometer (a) No. 47 (b) No. 48.	260
Fig. 10.31.	Piezometric head measured from piezometers and CPT holding tests at 11 months after surcharge.	261

Fig. 10.32.	OCR versus elevation interpreted from various tests.	262
Fig. 10.33.	Comparison of shear strength measured from various tests.	262
Fig. 10.34.	Comparison of effective stress measured from various tests.	263
Fig. 10.35.	Comparison of degree of consolidation interpreted from various tests.	263
Fig. 10.36.	Profile of soil at silt pond pilot area.	264
Fig. 10.37.	Construction sequence for pilot area and verification using data from silt pond pilot area.	266
Fig. 10.38.	Excess pore pressure versus time at large settlement area.	267
Fig. 10.39.	Excess pore pressure versus time for no-drain area.	267
Fig. 10.40.	Profile of soil at the area with the largest settlement in the silt pond.	268
Fig. 10.41.	Construction sequence for silt pond and verification using the data from the largest settlement area in the main work.	269
Fig. 10.42.	Profile of soil at the no-drain area.	271
Fig. 10.43.	Verification using data from the no-drain area in the main work.	272

LIST OF SYMBOLS

A	Area of sample
Al	Aluminium
b/r	Dimension less ratio and varied between 0 and 2
C	Particle concentration
Ca	Calcium
C_c	Compression index
C_c^*	Compression index in soil stage (Burland, 1990)
C_{c1}^*	Compression index in viscous stage
C_{c2}^*	Intrinsic compression index in second log cycle
C_{c3}^*	Burland's (1990) intrinsic compression index
CD	Chart datum
C_F	Large strain coefficient of consolidation
C_i	Compression index in ultra-soft soil stage
C_k	Hydraulic conductivity change index
c_v	Coefficient of consolidation
Cl	Chloride
Cr	Relative consistency
C_r	Recompression index
C_u	Undrained shear strength
CRL	Constant rate of loading test
CRS	Constant rate of strain test
CPT	Cone penetration test
CPTU	Cone penetration test with pore pressure measurement
D	Sample diameter
D	Double drainage (inward and outward)
D_{50}	Mean grain size

DS	Deep settlement gauge
D/H	Diameter to high ratio
d	Diameter of drainage filter
e	Void ratio
Δe	Void ratio change
e_0	Natural void ratio
e_f	Final void ratio
e_i	Initial void ratio
e_L	Void ratio at liquid limit
e_t	Void ratio at transition point
$e_{t(\Delta\sigma)}$	Transition void ratio which is a function of additional load
e^*_{10}	Void ratio at 10 kPa
e^*_{100}	Void ratio at 100 kPa
e^*_{1000}	Void ratio at 1000 kPa
e_{ts}	Void ratio at transition point based on settlement
e_{tp}	Void ratio at transition point based on pore pressure
Fe	Iron
FVT	Field vane test
f	Fluid
Gs	Specific gravity
G.D.S.	Geodetic digital system
H	Layer thickness
H_0	Initial thickness in soil stage
H_i	Initial thickness of ultra-soft slurry like soil
H_s	Thickness of soil at transition point
H_t	Thickness at time (t)
ΔH	Change in thickness (or) settlement
ΔH_i	Change in thickness in viscous stage
ΔH_{total}	Total settlement
h	Height of sediment
h	Head of water applied
I	Inward drainage
i	Hydraulic gradient
K	Potassium
k	Hydraulic conductivity
K_0	Coefficient of lateral earth pressure at rest
k_n	Hydraulic conductivity at certain void ratio
k_v	Hydraulic conductivity due to vertical flow
LL	Liquid limit

LI	Liquidity index
LVDT	Linear vertical displacement transducer
l	Thickness of sample
Mg	Magnesium
M_S	Total weight of soil
M_{WR}	Required quantity of water by weight to be added to obtain the required moister content
m	Numerical variable
m_v	Coefficient of volume compressibility
mCD	Meter above chart datum
N	Factor which related to thickness of sample
Na	Sodium
n	Porosity
n, m	Mathematical function
O	Outward drainage
O_{95}	Size of opening in a geotexile for which 95% of opening are smaller
OCR	Over consolidation ratio
OP	Open type piezometer
P_{atm}	Atmospheric pressure taken as 100 kPa
ΔP	Differential pressure
$\Delta P'$	Effective pressure change
PI	Plastic index
PL	Plastic limit
PP	Pneumatic piezometer
PZ	Vibrating wire electric piezometer
PWP	Pore water pressure
p	Pore pressure
p'_c	Preconsolidation pressure
p'_0	Overburden pressure
Q	Total volume of water drained out
ΔQ	Differential volume of water drained out
Q_0	Initial volume of water
q	Flow rate of water
S	Settlement
$S(T)$	Time of settlement
Si	Silicon
SP	Settlement plate
SO_4	Sulfhate

s	solid
T	Time factor
T_0	Time factor arbitrarily determined
T_F	Time factor for large strain consolidation
t	Time
t_i	Duration of delayed in pore pressure (min)
t_s	Time required to reach interception point in settlement measurement
t_p	Time when pore pressure dissipation commenced
Δt	Differential time
U_s	Degree of settlement
u	Settling rate
$\bar{u}$	Average pore pressure
u	Imposed excess pore pressure
u_i	Initial pore pressure
u_0	Excess pore pressure which may be exerted by the instantaneous loading of total pressure σ at T_0
u_b	Pore pressure at the base
Δu_b	Excess pore pressure at the base
VCL	Virgin Compression Line
VD	Vertical Drain
v	velocity
v_f	Velocity of fluid
v_s	Velocity of solid
W	Corrected moisture content
W_c	Weight of salt in water
WS	Water stand pipe
W_m	Measured water content
W_w	Weight of water
W_{w1}	Current water content
W_{w2}	Required water content
w	Weight of dry waste material
z	Solid thickness in material coordinate
z_0	Solid thickness in material coordinate at initial stage
z_1	Solid thickness in material coordinate at first stage
x, y, z	Coordinates
ρ	Density of water
ρ_s	Density of solid
ρ_f	Density of fluid

δe	Void ratio change
δh	Settlement
δp	Change in pore pressure
δt	Differential time
δu	Excess pore pressure
δv	Volume change
$\delta \sigma$	Total stress change
$\delta \sigma'$	Effective stress change
σ'_f	Final effective stress
σ'_i	Initial effective stress
σ'_v	vertical effective stress
σ'_t	Effective stress at transition point
σ	Total stress
$\Delta \sigma$	Additional stress
σ'	Effective stress
σ_γ	Stress contribution from the electrochemical forces
σ_i	Initial total stress
σ_v	Total vertical stress
ϵ	Power function which correlates with physical soil parameter
ϕ	Concentration of solid
ϕ_c	Critical concentration
ϕ_α	Concentration of solid when sedimentation is complete
α	Power function which correlates with physical soil parameter
α_v	Coefficient of volume compressibility
μ	Settling rate
ξ	Effective stress
λ	Coefficient
Δu_1	Pore pressure change
β	Power function which correlates with physical soil parameter
β	Interaction coefficient $(0-1)$
γ	Unit weight
γ_f	Unit weight of fluid
γ_s	Unit weight of soil
γ_w	Unit weight of water

ACKNOWLEDGEMENTS

Having travelled from one continent to another and receiving valuable guidances from many experienced mentors including learning under renowned professors and professionals in several tertiary institutions, the author wishes to acknowledge several people.

Firstly, the author would like to express his sincere gratitude to his pioneer teachers, the late Professor Ba Than Haq, retired Professor Dr Maung Thein and retired Professor Dr Maung Thinn of Yangon University, Myanmar for the excellent foundation laid for him in his geoscieneces career. The author would like to thank his mentors in his earlier career, U Sann Lwin, retired Director (Geotechnics), Irrigation Department, Myanmar; the late former Director (Projects), U Win Pe, Irrigation Department, Myanmar; former Mayor and Chairman, U Ko Lay of Yangon City Development Committee, Yangon, Myanmar, for training him in becoming a confident practicing engineer and geoscientist. Sincere appreciations are extended to retired Professor G. P. Jones of University College London, UK for his guidance during postgraduate study in hydrogeology. The author is also indebted to Professor Victor Choa and Associate Professor Wong Kai Sin of Nanyang Technological University, Singapore for their valuable guidances and kind encouragement throughout his research leading to doctoral degree in civil engineering.

The author would like to thank his mentors Dr. A Vijiaratnam, the former Chairman of SPECS Consultants Pte. Ltd., Singapore and Mr. Law Kok Hwa the former Senior Vice President of PSA Corporation, Singapore for their guidances and support throughout his career.

Special thanks are also extended to Professor Richard Jardine of Imperial College, London, UK; Dr. David Hight, Director, Geotechnical Consulting Group, London, UK; Professor Serge Leroueil, Laval University, Canada; Professor A. S. Balasubramaniam, Griffith University, Australia;

Associate Professor Hamid Nikraz of Curtin University, Australia; Dr. A. Arulrajah of Swinburne University of Technology, Australia; Associate Professor J. Chu of Nanyang Technlogical University and Dr. M. F. Chang, Associate, Shannon & Wilson Inc. USA; Dr. Ho Kong Meng, Senior Executive Engineer, PSA Corporation, Singapore, Mr K. H. Loh, Managing Director of Loh & Loh Construction; Dr. T. Sivapatham, Director Geonamics; Mr. C. P. Seh, General Manager, Kisho-Jiban Ltd., Singapore; Dr. Y. M. Na, Chief Engineer, Hyundai Engineering and Construction, South Korea; for their supports and co-operation given throughout his journey in industrial practice and research.

Finally but not least, the author would like to express his appreciation to his wife Win Myint Than, daughters Htet Ei Bo, Thanda Bo and son Kyee Soe Bo for the understanding and support given to him during the preparation of this book and throughout his career.

Dr. M. W. BO (BO Myint Win)

ABOUT THE AUTHOR

Dr. M W BO (BO Myint Win) is a Director (Geo-Services) in DST Consulting Engineers Inc. Canada. He graduated with B.Sc (Geology) from the University of Rangoon, Myanmar and received Postgraduate Diploma in Hydrogeology from University College London, UK and MSc Degree from University of London, UK. He obtained his Ph.D in Civil Engineering (specialized in geotechnics) from the Nanyang Technological University, Singapore. He is a Fellow of the Geological Society, London, UK and a Fellow of Institution of Civil Engineers, UK. He is also a professional geologist, chartered Geologist, Chartered Scientist, Chartered Engineer, Chartered Environmentalist, European Geologist and European Engineer. Dr. Bo is a member of panel of experts for Federation of European Geologists and also a committee member of Technical Committee 39 for International Society of Soil mechanics and Geotechnical Engineering, committee member of Technical Committee for International Geosythetic Society and also a committee member of Soil and Rock committee D-18 for American Society of Material Testing. He worked for the Myanmar Government as an Engineering Geologist cum Hydrogeologist for 11 years and a Water Resources Engineer for 2 years. He worked in Singapore as a Consulting Geotechnical Engineer for 11 years in SPECS Consultants Pte Ltd and headed the Geotechnical Division in the Changi East Reclamation Projects in Singapore. He also acted as a Consultant and Technical Advisor for some reclamation and ground improvement projects in the Far East. Dr. Bo worked in an Faber Maunsell Ltd, UK as a Technical Director for 4 years till mid 2007. Dr. Bo is an experience practicing engineer as well as educator and he has been giving several special lectures and workshops in the international conferences, tertiary institutions and professional associations. He is a co-supervisor of a few Ph.D students in Swinburne University of Technology, Australia and also an external examiner of higher research degrees at Swinburne University of Technology, Australia. He is also a research partner with Swinburne University of Technology for a government funded research projects. He has published over 100 papers in international journals and conferences. He is also the first author of textbooks entitled *"Soil Improvement"* and *"Reclamation and Ground Improvement"*. He has also written a chapter in the textbook, *"Ground Improvement, Case Studies"*. Due to his significant contributions to the engineering industry and extensive industrial research works carried out while he was working as consultants in the Far East, his name has appeared in *Who is Who in The World* 2007 & 2008 and also was appeared in *Who is Who in Science and Engineering* 2008.

Chapter 1

INTRODUCTION

1.1. Deformation of Soils and Their Compressibility

There are various types of deformation of soils such as plastic flow, elastic deformation, shear deformation, undrained creep, primary compression, secondary compression, and liquefaction. Most theories of deformation and compression cover deformation of either cohesive or cohesionless soils but there are some theories which cover cohesive and frictional soils. Some types of deformation are stress-dependent and some are time-dependent. Generally, majority of large strain deformations are associated with cohesive compressible soils. In 1923, Terzaghi proposed a time-independent linear-elastic model of compression behavior for a low permeable thin layer of soil. Taylor and Merchant (1940) explained the compression behavior after the primary stage with an elastoviscous model. With these two models, the magnitude and time rate of compression of a soil in the primary and secondary stages can be predicted.

However, studies on compressibility of soils by the early researchers concentrated mainly on naturally sedimented clay deposits, which are either normally consolidated or over-consolidated. On the other hand, some researchers in the chemical engineering field such as Coe and Clevenger (1916), Kynch (1952), Fitch (1966a, b), and Bustos (1987) are among others who studied the sedimentation process. McRoberts and Nixon (1976), Been and Sills (1981), and Tan (1995) applied the sedimentation theory to soil sedimentation.

Tan *et al.* (1988) studied the transition of sedimentation process to consolidation process. Some researchers such as Been and Sills (1981), Lee and Sills (1981), Mikasa (1961), and Toorman (1996) investigated the self-weight consolidation process. Consolidation or compression of slurry-like soil was studied by Carrier III and Beckman (1984), Carrier III and Keshian (1979), and Carrier III *et al.* (1983). Tan *et al.* (1988) has proposed a nonlinear equation for predicting hydraulic conductivity and effective stress from void ratio for stress level of up to 50 kPa. However, the deformation behavior of ultra-soft slurry-like soil upon additional load is not well understood.

Due to the high demand of land in coastal areas, reclamation or remediation of land on ultra-soft soil like recently deposited deltaic or estuarine deposits, waste ponds, and mine tailing ponds became necessary. Such deposits are extremely soft and have very high water content. In addition, these materials are underconsolidated and usually still undergoing self-weight consolidation.

The compression and consolidation behavior of the ultra-soft soil due to additional load is different from that of the normally or overconsolidated soil. The prediction of magnitude and time rate of deformation using Terzaghi's consolidation theory may not be appropriate (Bo *et al.*, 1997a, c and 1999). Terzaghi's consolidation theory leads to an underestimation of the magnitude and an overestimation of degree of consolidation.

Therefore, the settlement prediction during and after reclamation on this type of soil deposit is extremely difficult due to a lack of a deformation model. As such planning and scheduling of soil improvement works become impossible since the duration required for consolidation is not known. Therefore, the development of a compression model which can predict the compression behavior during the viscous stage is important and necessary. Been and Sills (1981) have suggested that a suitable theoretical model for very soft soil is required to be developed.

1.2. Related Previous Research

Past researchers concentrated mainly on the compressibility of normally or overconsolidated soils. Terzaghi's theory was originally developed for a thin layer of compressible soil, and is not easily adaptable for a thick layer of soil involving large strains. Subsequently Gibson and Lo (1961) proposed a model similar to that proposed by Taylor which took into consideration of Large strains by applying a Lagrangian coordinate system.

Since the compression and consolidation of soil is nonlinear, Mikasa (1961) proposed a nonlinear consolidation model using volumetric strain and specific volume rather than settlement and void ratio. Mikasa and Takada (1995) derived a new consolidation theory which accounts for the continuous changes in the consolidation properties with depth. Special consideration for the effect of self-weight of clay and the reduction of thickness were taken into account in addition to the change in soil parameters with respect to space and time.

Since nonlinearity alone is not sufficient to model the compression and consolidation behavior of a soil, viscous effect was taken into consideration by the models of Barden and Berry (1965), Garlanger (1972), Mesri and Rokshar (1974), Leroueil (1994), and Yoshikuni et al. (1994).

However, the viscous effect considered in their models is either in the primary consolidation stage or after the primary consolidation stage.

The vertical deformation due to vertical and radial flows has been proposed by Carillo (1942) and Barron (1948). A three-dimensional consolidation model was proposed by Biot (1955) which takes into account the volumetric strain and the all-round flow of pore water from the soil mass.

Coe and Clevenger (1916), Kynch (1952), and Fitch (1966a, b, 1975 and 1979) are among others who had studied and explained the sedimentation process in their studies of the sedimentation of thickeners. McRoberts and Nixon (1976) and Been and Sills (1981) applied the sedimentation theory to soil sedimentation and carried out laboratory sedimentation experiments together with the measurement of small effective stress gain.

Beyond sedimentation, researchers have been studying self-weight consolidation. Been and Sills (1981), Lee and Sills (1981), Mikasa (1961), Tan et al. (1988), and Toorman (1996) were among others who have studied this process. Carrier III and Beckman (1984), Carrier III and Keshian (1979), Carrier III et al. (1983), Carrier III and Bromwell (1980) widely studied the compression of slurry-like soil and remediation of slurry waste pond. Determination of consolidation properties of very soft clay was studied by Tan et al. (1988) with the measurement of pore pressure and density with Gamma rays. Prediction of consolidation of very soft soil was reported by Cargill (1982a, b) using various time factor charts derived from large strain consolidation theory. The consolidation of soil stratum, including self-weight effects and large strain derived from Gibson et al. (1967) was proposed again by Lee and Sills (1981). However, Salem and Krizek (1973) used conventional theory and method to characterize the dredged slurries. The area that is less studied is that of a soil subjected to a surcharge while undergoing self-weight consolidation. In addition to the models and time factor curves required to predict the compression of ultra-soft soil, some test methods such as constant rate of loading, constant rate of strain tests on very soft soil or slurry were carried out by Carrier III and Beckman (1984). A more detail review of their research will be discussed in the relevant subsequent chapters.

1.3. Outline of the Book

This book aims to explain the compression and consolidation behavior of an ultra-soft soil subjected to an applied load. This soil in question was still undergoing self-weight consolidation. The magnitude and rate of settlement during the early stages of compression and consolidation were explained using various experimental tests carried out by an author during the late 1990s. Subsequently, a numerical model is developed by the author for the prediction of magnitude and time rate of deformation of ultra-soft soil during the slurry stage.

The book describes the experimental laboratory and field tests carried out by the author and his coworkers to investigate the compressibility and consolidation of ultra-soft soil. Test results from such comprehensive laboratory tests carried out on samples with various moisture contents and different applied loads were presented and discussed extensively in this book. Results from tests carried out with both small-scale and large-scale consolidometers are widely discussed in Chapters 5 and 6. Both consolidometers used by the author were equipped with pore pressure transducers and settlement monitoring devices. With these tests, the compression and consolidation behavior during the viscous stage were explained. How to determine the transition point from the slurry stage to the soil stages applying these types of tests is described in Chapters 5, 6, and 7. The parameters controlling the magnitude and the rate of deformation are explained and discussed. Consolidation tests carried out with Rowe cell allowed to find out the effect of flow pattern on the deformation behavior. Based on several combinations of Rowe cell tests carried out by the author and his coworkers, a basic equation for the prediction of magnitude of settlement as well as the relationship of various compression indices with the void ratio at liquid limits are proposed. The relationship between void ratio and hydraulic conductivity, time factor curves for prediction of time rate of settlement are also proposed. Several constant rates of loading tests and constant rates of strain tests are also presented to explain the relationship between the void ratio change and effective stress gain. Based on the results from the experimental study extensively carried out by the author, a simplified mathematical model has been adopted. Compression parameters and a large strain coefficient of consolidation parameters are proposed for prediction of the rate and magnitude of compression of ultra-soft soil with this model. The validation of the proposed model is made against the laboratory-controlled measurements. Comparisons between the proposed model and existing models are made and discussed.

In addition to the experimental studies in the laboratory, field tests carried out are also described. The compression and consolidation behavior in the field are monitored by piezometers and settlement gauges throughout the process, and the relationship between the gain in effective stress and deformation is explained. The verification of the improvement of ultra-soft slurry-like soil during and after deformation by various *in situ* testing and laboratory methods is also presented. Finally, the reliability of the proposed model is verified against laboratory results and field data, especially the measured settlements. Validations are made for soil settlement under two conditions: with and without vertical drains.

Chapter 2

SEDIMENTATION AND CONSOLIDATION

2.1. Sedimentation

The formation of soil typically goes through two stages. The first stage known as sedimentation involves the conversion of discrete soil particles in a suspension into loose sediment. The sedimentation velocity of clay suspensions represents a statistical average of all solid particles that is governed by the combined action of gravitational force, Brownian force, interparticle electrical force, Van der Waals force, and Stokesian viscous force (Russel et al., 1989). The theory of gravitational settling of incompressible solids was developed by Kynch (1952) in the chemical engineering field and was subsequently applied to soil sedimentation by McRoberts and Nixon (1976). The second stage is the self-weight consolidation where the process starts at the bottom while the sedimentation process continues at the top.

The sedimentation process has been extensively studied, starting with the work of Mischler (1912) and Coe and Clevenger (1916). Their methods have been used in the design of thickeners in the chemical engineering field. In 1952, Kynch presented the first well-known theory of sedimentation. His theory is purely based on kinematics. It describes the sedimentation under gravity of solid particles through fluid as a wave propagation phenomenon. Kynch established that in such a process, discontinuities might appear as shock waves or contact discontinuities. He assumed that the velocity of the solid particles in a fluid depends only on the local concentration. No wall effect was taken into account, so that motion is one-dimensional and the particles are assumed to be of the same size and shape. In his theory the settling rate u is expressed as

$$u = u(C) \tag{2.1}$$

where C is the particle concentration in the suspension fluid.

Kynch's assumption is not valid throughout the column in which a zone of compacting sediment forms at the bottom. In the compaction zone itself, u is not dependent on C alone, but also on the solid stress gradient (Michaels and Bolger, 1962; Fitch, 1966, 1975, 1979; Shirato et al., 1970). Just above the compaction zone, the Kynch zones or characteristics do not arise from the origin (i.e. base of settling column), as required by Kynch's Theory, but arise from the interface between the compacting sediment and the zone settling suspension above it (Fitch, 1966, 1979; Tiller, 1981).

The first well-known modification of the theory was by Talmage and Fitch (1955) and discussed by Scott (1966) and Fitch (1966, 1979).

In Fitch's modification, three modes or types of sedimentation were separated: (i) **clarification** in which flocculates are separated and settled independently; (ii) **zone settling** in which flocculates are incorporated into some solid structure so that they are all constrained to subside at more or less the same rate; and (iii) **compression or compaction** in which the solid structure is strong enough to exhibit a compressive yield value.

These three types of sedimentation are illustrated in Fig. 2.1. The left side represents particles with little tendency to cohere, the right side represents those for which the interparticle cohesion is strong. The vertical axis represents particle concentration, with more concentrated suspension at the top.

At low concentration, a regime called clarification occurred. The particles are, on the average, far apart and able to settle independently, but collisions occur. If the particles then cohere, they grow into clumps or flocculates. The settling rate increases as they grow. On the other hand, if they do not cohere, each moves downward at a characteristic rate. Therefore, there are two types of settling in clarification such as flocculent and particulate. There is no sharp boundary between these two, as one grade will gradually move to the other.

When the particles become more concentrated, crowded, and closer together, they finally reach a point where each is in contact with the others. If they then have any tendency to cohere, they link into some sort of a flocculated structure. Whether the structure formed is continuous (Fitch, 1962, 1972) or consists of a bed of closely spaced flocculates (Roberts, 1949) is still in question. However, in soil or slurry it is well

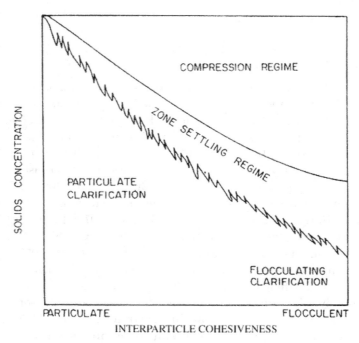

Fig. 2.1. Regimes of sedimentation (after Fitch, 1979).

known that the beds are not continuous and are formed with significant voids. These voids are either filled with water or air. But in any case, particles held in the structure are constrained to settle at the same rate. The solids subside with a sharp interface between solid and suspension. The slurry enters a zone-settling regime and exhibits line-settling behavior, in which a visible interface moves downward from the top, rather than building upward from the bottom as in clarification.

As the concentration is further increased, the pulp structure becomes so firm that it begins to develop strength. Each layer of solids is able to provide mechanical support to the layers above. Under such a condition, the pulp is said to be in compression and the regime is called compression or compaction. All three regimes may not be present in any continuous sedimentation pool. But if they were, they would ideally be distributed as shown in Fig. 2.2.

According to Coe and Clevenger (1916), a critical zone is conceived to form in the zone-settling regime. Fitch (1966) and Dixon (1977a, b) contend that it can also be formed in the compression regime.

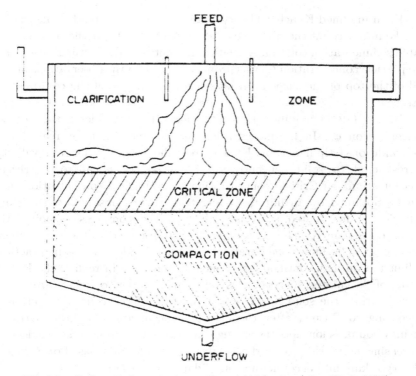

Fig. 2.2. Sedimentation zone in continuous thickening (after Fitch, 1979).

After Fitch (1962, 1966, 1979) modified Kynch's Theory of Sedimentation by taking into consideration the compaction zone, Tiller (1981) further modified the theory by taking into account the upward movements of the compaction boundary. According to Tiller (1981), Kynch's procedures involve fundamental errors, as they do not take the rising sediment into proper account. Ignoring the appearance of the bottom sediment, Kynch postulated that sedimentation velocity (settling rate), u, was a unique function of solid concentration C, and derived a simple, first-order partial differential equation describing the behavior of the settling particles.

Tiller (1981) proposed that Kynch's Theory can be corrected or extended to cover the case in which a subjacent compression zone is formed. This is done by taking into consideration the rise of the suspension–sediment interface, as well as the fall of the suspension–supernatant interface.

Fitch modified Kynch's Theory again in 1983. He extended the theory to take into account the presence of a compaction zone at the bottom of a batch sedimentation column. A relationship between the settling rate u and the particle concentration C, was proposed based on the observations of the fall of the top of the suspension interface and the rise of the compression interface.

Bustos (1987) modified Kynch's Theory by introducing the critical concentration ϕ_c. In his modification, he assumed that for real slurry, all conditions stipulated by Kynch are valid, except that the validity is restricted to solid concentration less than or equal to a critical concentration, ϕ_c. Critical concentration is the highest concentration that can be obtained by an initial settling test. Bustos observed the settling behavior of flocculated suspension beyond the compression point. He explained that in Kynch's Theory, settling terminated at the point where $\phi_c = \phi_\alpha$ (ϕ_α is final concentration in Kynch model) where in practice, sedimentation may continue for a much longer time. In reality, the Kynch model of flocculated suspension is valid only for $\phi \leq \phi_c$, since after that concentration, the solid flux is not a function of the concentration only. According to Bustos, after the critical concentration, ϕ_c, the particles go into compression and the change in height of sediment still exists in decreasing slope. He proposed a height versus time plot based on Kynch's Theory taking into account the consolidation effect (Fig. 2.3).

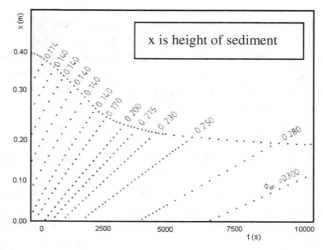

Fig. 2.3. Sedimentation which takes into account the consolidation (after Bustos, 1987).

2.2. Sedimentation to Consolidation

As explained in the earlier section, for a given type of slurry, there is no sharp boundary of time between sedimentation and consolidation. These two processes occur concurrently, i.e. while sedimentation takes place at the top of the slurry column, the sediments at the bottom commence consolidating. At that time, a boundary between sediments and slurry suspension will be formed. This boundary will rise due to subsequent sedimentation of particles from the suspension slurry.

While the boundary of sediments is rising, the bottom part of the sediments starts consolidating under self-weight. At that time, the sediment behaves like a soil governed by Terzaghi's principle of effective stress. For a one-dimensional two-phase flow, the governing equations in Eulerian coordinates assuming incompressible fluid and solid phases have been presented by Tan et al. (1990a) as follows:

(A) Continuity Equations

Solid Phase

$$\frac{\delta(1-e)}{\delta t} + \frac{\delta(1-e)v_s}{\delta x} = 0 \tag{2.2}$$

Fluid Phase

$$\frac{\delta e}{\delta t} + \frac{\delta e v_f}{\delta x} = 0 \tag{2.3}$$

(B) Equations of Equilibrium

$$\frac{\delta \sigma'}{\delta x} + \frac{\delta u}{\delta x} + \gamma_s(1-e) + \gamma_f e = 0 \tag{2.4}$$

(C) Darcy's Law (Equation of motion for fluid)

$$e(v_f - v_s) = -\frac{k}{\gamma_f}\left(\frac{\delta u}{\delta x} + \gamma_f\right) \tag{2.5}$$

where the subscripts f and s refer to the fluid and solid phases respectively, e is the void ratio, v is the velocity, γ is the unit weight, u is the pore pressure, σ' is the effective stress, and k is the hydraulic conductivity.

From sedimentation to consolidation, the major argument is whether there is a sharp boundary between sedimentation and consolidation. According to Imai (1981), and Been and Sills (1981), there is no sharp boundary, instead there is a transition zone between the zone where sedimentation dominates and the zone where consolidation dominates. However, Li and Williams (1995) claimed that during the early stages, the density of the upper zone was quite different from that of the lower zone and there existed between them a very narrow transition band with sharp boundaries. In reality, if both sedimentation and self-weight consolidation are in progress, a sharp boundary may not be formed and the change shall be gradual.

Been and Sills (1981) and Bowden (1988) attempted to determine the transition point of slurry to soil by measuring the difference between total stress and pore pressure. This concept was illustrated by Hight et al. (1987) using an idealized transition point between suspension and soil (Fig. 2.4). However, Tan (1995) argued that Terzaghi's principle of effective stress requires an effective stress that not only exists, but also controls the deformation. He suggested that the transition point should be the point at which effective stress becomes valid.

Tan (1995) again stated that there is a gap between the void ratio when effective stress is first measured and the void ratio at the surface at the end of self-weight consolidation. Based on Been and Sills' (1981) results, the typical value of void ratio of the surface at the end of all settling is usually around 6 and 7 while the void ratio when the effective stress is first measured is usually around 9 and 10.

Tan (1995) also explained that in the initial stage when the clay particles come close, there is a chemical bonding as a result of the electrolytic condition and Van der Waals forces. The initial presence of this interactive stress may not imply the formation of a Terzaghi's soil. Therefore, Tan suggested that it would be better to think of effective stress which controls the deformation, and that it may not be exactly equal to the difference between the total stress and pore pressure. He has proposed a more appropriate definition of effective stress as follows:

$$\sigma' = \sigma - u - \sigma_\gamma(t) \qquad (2.6)$$

where σ_γ is the contribution from the electrochemical forces that has no direct effect on the deformation and may be time-dependent.

When the particles come close enough to interact, σ_γ develops and is equal to $(\sigma - u)$, and there is no effective stress. This phenomenon explained the discrepancy between the two density values measured at the end of settling and when the difference between total stress and pore pressure is first measured. This is because one is measured when σ_γ first appeared

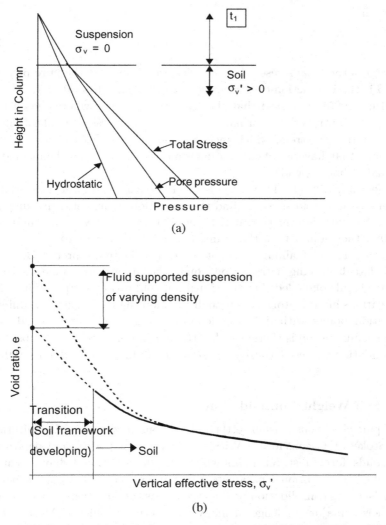

Fig. 2.4. Idealized transition point between suspension and soil (after Hight et al., 1987).

and the other is measured when σ' first appeared. Generally, the density at the surface of the column after settling has ceased is much higher than the density when a difference between total stress and effective stress is first measured.

Pane and Schiffman (1985) proposed that there is a transition stage in which a partial effective stress is present. They suggested the following equation:

$$\sigma = \beta(e)\sigma' + u \qquad (2.7)$$

where σ is the total stress, σ' is the effective stress, u is the pore pressure, and β is the interaction coefficient which will transit from 0 to 1.

Tan (1995) suggested that the transition point is where the effective stress first controls the deformation; this effective stress is smaller than the total interactive stress, which includes contribution from electrochemical reactions that have no effects on deformation. This electrochemical stress is usually time-dependent.

Been and Sills (1981) and Tan et al. (1988) carried out the measurement of density and yield stress to find the transition point between slurry and soil. Although they investigated different types of clay with liquid limit of 80%, they found that the transition point for slurry to soil occurred at a void ratio of about 6. Hight et al. (1987) reinterpreted Been and Sills' data by taking a transition point when relatively stable void ratio occurred and concluded that the transition void ratio was approximately 4. The various investigators were actually attempting to measure the different transition points with different definitions. The void ratio of sediments after sedimentation is therefore of particular importance in the subsequent consolidation process (Abu-Hejleh et al., 1996; Liu and Znidarcic, 1991).

2.3. Self-Weight Consolidation

Clay particles in suspension settle under a combined action of gravitational and Stokesian viscous forces. Within a certain time period after deposition, the solids form a structure in which the deformation behavior can be described by traditional parameters in soil mechanics. For such a period, there is a long consolidation process termed as the self-weight consolidation which is a major challenge in many engineering problems (Mcvay et al., 1986; Huerta et al., 1988).

When self-weight consolidation commences, it becomes possible to measure the effective stress which eventually increases equivalent to the effective overburden stress when the self-weight consolidation process is completed.

Been and Sills (1981) and Li and Williams (1995) claimed that for slurry with low initial solid concentration the self-weight consolidation starts at the bottom of the slurry and proceeds upward.

Imai (1981) showed that the self-weight consolidation occurred in the lower part of a settling column simultaneously with sedimentation in the upper part of the column. However, Scotts et al. (1986) claimed that the sedimentation and consolidation processes are distinct. They stated that during sedimentation, the slurry density does not change except at the bottom of the settling column. After sedimentation is completed, consolidation proceeds downward from the surface of the slurry. These two hypotheses contradict each other. One states that since the material at the bottom of the column has the highest self-overburden weight, it helps to squeeze out the water. Therefore, consolidation started at the base. However, the other authors suggested that the consolidation process starts near the drainage face which is at the top.

Li and Williams (1995) stated that if the slurry is less than or equal to 35% solid concentration by mass, the effective stress develops from base upward. This means self-weight consolidation starts at the base and proceeds upward. They stated that this is not the case for slurry with a high initial solid concentration in which effective stress is detected simultaneously at all positions.

Li and Williams (1995) carried out sedimentation and self-weight consolidation tests with pore pressure measurement in a column and reported that the effective stress develops at the base and proceeds upward. With these test results, they claimed that the self-weight consolidation started at the base and proceeded upward, due to the formation of the sediment first at the base.

However, based on one of their tests carried out with solid concentration of 40%, the results did not support the above concept. The measured effective stress did not increase with depth and there were some local effective stresses higher than the effective stress near the base (Fig. 2.5), in which effective stress is equal to $(\sigma - u)$. This is consistent with the finding of Kearsey and Gills (1963) on the development of local high effective stress.

However, based on the moisture content and current effective stress measurement at the slurry pond (siltpond) in Changi East, Singapore, it was confirmed that the self-weight consolidation started at the base.

2.4. Theory of Self-Weight Consolidation

Been and Sills (1981) explained that soil particles that are laid down by sedimentation through water form a loose structure, the stiffness of which

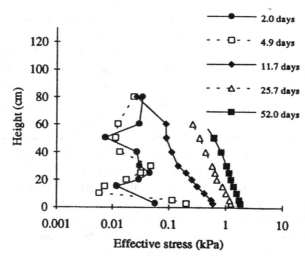

Fig. 2.5. Profile of effective stress increment with time from sedimentation test on high initial solid concentration slurry (solid concentration = 40%) (after Li and Williams, 1995).

then gradually increases with additional load. In that case, consolidation is caused by the self-weight of the soil and accompanied by large strains. Been and Sills (1981) described that the traditional soil consolidation theories are inadequate to explain the consolidation due to the self-weight body forces. They carried out sedimentation and self-weight consolidation tests in the laboratory. In their study, they measured the profile of the slurry density and also measured the pore pressure with depth. By measuring those parameters, they managed to determine the change in density along the profile as well as the effective stress profile. Based on their study, they proposed the following governing equation for self-weight consolidation:

$$e(y, T) = e_i - \beta z_0 \times \left[1 - y - 2 \sum_n \frac{\cos m\pi y}{m^2 \pi^2} \exp(-m^2 \pi^2 T) \right] \quad (2.8)$$

where $z =$ solid thickness or material coordinates,
$y = z/z_0$
$T = C_F t/z_0^2$ is the time factor,
$C_F =$ coefficient of large strain consolidation,
$e =$ void ratio,
$e_i =$ initial void ratio,
$\beta = (\rho_s - \rho_f)/\alpha$,

ρ_s = density of solid,
ρ_f = density of fluid,
α = constant
$n = 0, 1, 2,$
$m = 1/2(2n+1).$

Isochrones of void ratio worked out from Eq. (2.8) is shown in Fig. 2.6. The time rate of self-weight consolidation is defined as

$$\frac{\delta e}{\delta T} = \frac{\delta^2 e}{\delta y^2} \quad \text{for } 0 \leq y \leq 1 \text{ and } T \geq 0 \tag{2.9}$$

The excess pore pressure distribution for soil plus the imaginary overburden is given by Eq. (2.10).

$$u(y,T) = 2(\rho_s - \rho_f)z_0 \times \sum_n \frac{\cos m\pi y}{m^2 \pi^2} \exp(-m^2\pi^2 T) \tag{2.10}$$

Boundary conditions and the definition of the modified problem are shown in Fig. 2.7. The excess pore pressure isochrones worked out from Eq. (2.10) is shown in Figs. 2.8 and 2.9. The self-weight consolidation model by Been and Sills (1981) was supported by case studies for some soils with various initial densities for various duration of settling. Comparisons of the theoretical model with measured surface settlements and densities for various duration are shown in Figs. 2.10(a)–2.10(f). It was found that Been and Sills (1981) closely predicted the surface settlement during sedimentation as well as self-weight consolidation stages.

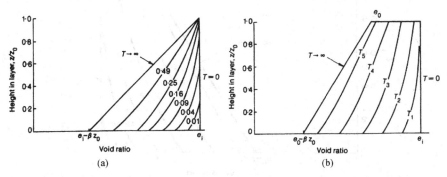

Fig. 2.6. Isochrones of void ratio: (a) after Lee (1979) or Eq. (2.8); (b) expected from experiments (Been and Sills, 1981).

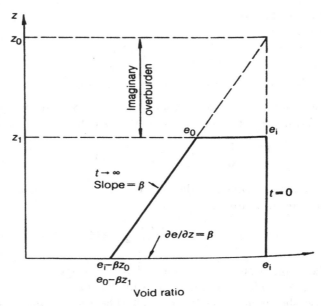

Fig. 2.7. Boundary conditions and definition of the modified problem (after Been and Sills, 1981).

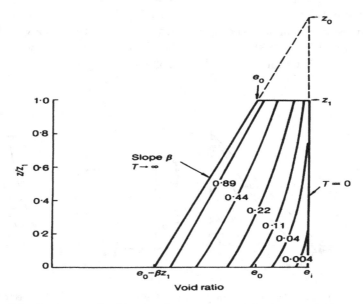

Fig. 2.8. Isochrones of void ratio when $z_0 = 1.5z_1$ (after Been and Sills, 1981).

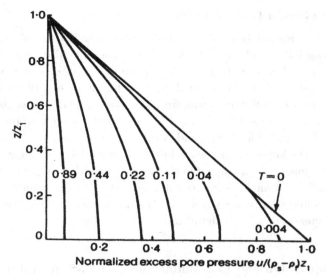

Fig. 2.9. Isochrones of excess pore water pressure when $z_0 = 1.5 z_1$ (after Been and Sills, 1981).

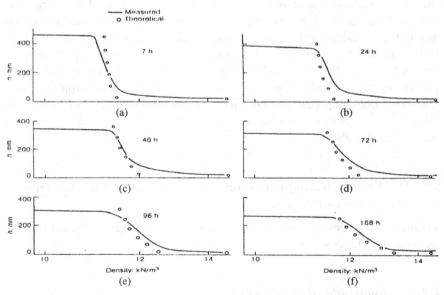

Fig. 2.10. Comparison of theoretical model with surface settlements and real densities for various duration (after Been and Sills, 1981).

2.5. Large Strain Consolidation

Large strain consolidations are generally associated with thick layers of compressible soil usually with high moisture content. Terzaghi's theory is unsatisfactory when the strains involved are substantial (order of 20% or more) or when self-weight is involved (Schiffman et al., 1984). Solution for large strain consolidation was first proposed by Gibson et al. (1967). Details of their proposal would be discussed in the later chapters. Cargill (1982a, b), Lee and Sills (1981), and Been and Sills (1981) are among others who applied the large strain consolidation theory to very soft soil including self-weight consolidation stage. On the other hand, Mikasa (1961) proposed another method of analysis for consolidation of soft clay including the self-weight consolidation process based on a similar theory. This study would also be discussed in greater detail in a later chapter.

2.6. Compression and Consolidation of Ultra-Soft Soil under Additional Load

Although self-weight consolidation and large strain consolidation were studied by various researchers, the deformation behavior of an ultra-soft soil which is undergoing self-weight consolidation upon additional load is still not yet understood. This behavior is likely to involve large strains. Since significant additional load is applied, the contribution of self-weight consolidation to the computation of settlement is believed to be negligible. Therefore, large strain consolidation alone may be applicable in this case. This book explains and describes the compression and consolidation behavior of an ultra-soft soil upon additional load with various laboratory experimental tests.

2.7. Prediction of Magnitude and Rate of Settlement in Large Strain Consolidation

Various researchers have attempted to predict both magnitude and rate of settlement in large strain consolidation.

2.7.1. *Magnitude of Settlement*

The prediction of ultimate settlement for very soft soil or slurry has been proposed by various researchers. Salem and Krizek (1973) applied the conventional Terzaghi's infinitesimal strain theory and used conventional parameters to predict the consolidation of dredged slurries. Their solutions are however not realistic to be applied in practice.

Tan et al. (1988) carried out large strain consolidation tests with pore pressure, density, and total stress measurements using pore pressure transducers, Gamma rays, and total pressure cells, respectively. They proposed a relationship between void ratio and effective stress up to 200 kPa, which gave a good correlation with their data between 0.5 and 20 kPa (Fig. 2.11). However, their tests were carried out only in a narrow range of void ratio between 5.8 and 6.7. Their void ratio-effective stress relationship therefore still shows a single relationship curve below 20 kPa. There should in fact be several curves in low stress range dependent upon the initial moisture content as reported by Leonards and Ramiah (1960), Olson and Mitronovas (1962), Olson and Mesri (1970), Woo et al. (1977), Carrier III and Bromwell (1983), and the tests carried out by the

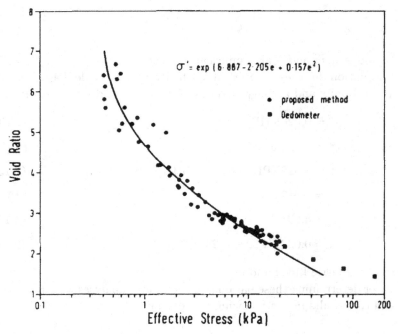

Fig. 2.11. Void ratio-effective stress relationship for very soft clay (after Tan et al., 1988).

author. Gibson et al. (1967) also proposed an e-log σ' relationship to be able to predict the magnitude of settlement. Details of these will be discussed in a later chapter.

2.7.2. Time Rate of Settlement

The researchers who studied the magnitude of settlement also studied the time rate of settlement. They developed finite difference or finite element programs, applying the finite strain theory by Gibson et al. (1967). Carrier III and Beckman (1984) used a numerical model which required effective stress-void ratio-hydraulic conductivity relationships expressed by power functions (Carrier III and Bromwell, 1983; Carrier III and Beckman, 1984). They proposed a void ratio, e versus effective stress, σ' relationship.

$$e = \alpha \left(\frac{\sigma'}{P_{\text{atm}}}\right)^\beta + \varepsilon \qquad (2.11)$$

and for hydraulic conductivity,

$$k = \mu \frac{(e-\lambda)^\nu}{1+e} \qquad (2.12)$$

where $\alpha, \beta, \varepsilon, \mu, \nu$, and λ are empirical coefficients.

Estimation of these empirical coefficients was made based on the Atterberg Limits and activity of slurry (Carrier III and Beckman, 1984).

$$\alpha = 0.0208 \text{ (PI) } [1.192 + (\text{act})^{-1}] \qquad (2.13\text{a})$$

$$\beta = -0.143 \qquad (2.13\text{b})$$

$$\varepsilon = 0.027 \text{ (PL)} - 0.0133 \text{ (PI) } [1.192 + (\text{act})^{-2}] \qquad (2.13\text{c})$$

$$\mu = 0.0174 \text{ (PI)}^{-4.29} \text{ m/s} \qquad (2.13\text{d})$$

$$\nu = 4.29 \qquad (2.13\text{e})$$

$$\lambda = 0.027 \left[(\text{PL}) - 0.242 \text{ (PI)}\right] \qquad (2.13\text{f})$$

where PI = Plastic Index and act = activity.

After determining these parameters, computer analyses were performed to yield the results in the following expression:

$$h = a\left(\frac{w}{A}\right)^b t^{c-b} \qquad (2.14)$$

where
- h = height of sediment;
- w = weight of dry waste material;
- A = size of disposal area;
- t = elapsed time; and
- a, b, c = analytic coefficients determined by making several computer runs that bracket the range of expected filling rates.

However, this method cannot be used in the conventional way of predicting time rate of settlement using the familiar time factor curves. Gibson *et al.* (1981), Lee and Sills (1981), and Cargill (1982a, b) also proposed the time factor curves and equations to predict the time rate of settlement for large strain consolidation. The details will be discussed in a later chapter.

Although the whole process of sedimentation to consolidation was discussed in the literature review, this book will only be emphasized on the 1D compression of ultra-soft soil upon additional load. The explanation will be made on the basis of the laboratory and field measurement data.

Chapter 3

MODELS AND ANALOGY

Consolidation models of material have been applied to solve the soil compression and consolidation behavior. Since saturated soil has two phases such as water and solid, the model used consists of viscous model which follows fluid mechanics and elastoplastic model which follows solid mechanics. Therefore, there are some models, which are combination of fluid characteristics and solid characteristics. Initiated by Terzaghi's linear elastic model (1923) followed by Taylor and Merchant (1940) linear elastoviscous model, Gibson and Lo (1961) linear elastoviscous model, Barden and Berry (1965) non-linear elastoviscous model are among the models with combined viscous flow and elastic deformation of soil behavior. Since soil consists of not only elastic behavior, but also plastic behavior, Garlanger in 1972 introduced the elastoviscous plastic model. Summary of existing soil models is shown in Fig. 3.1.

3.1. Viscoplastic Model for Ultra-Soft Soil

Since ultra-soft soils are in a state of under consolidation, the slurry-like soil behavior can be explained by a combined viscous and plastic deformations. There could be two scenarios of deformation. In the first scenario, the ultra-soft soil will start with viscous deformation and then switch to plastic deformation. The transition is likely to be smooth. This model can be explained by a spring and dashpot model shown in Fig. 3.2(a) in which viscous deformation is represented by a dashpot and plastic deformation by a spring. The viscous deformation is likely to be time-dependent and irreversible. Both viscous and plastic deformations may be linear in its own

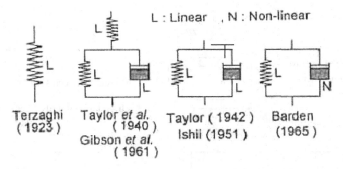

Fig. 3.1. Summary of existing soil models.

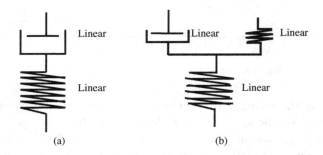

Fig. 3.2. Viscoplastic model for ultra-soft soil.

stage. In the second scenario, the deformation of slurry-like viscous material starts with viscoplastic behavior and at a certain point it smoothly switches to plastic deformation. This second scenario is illustrated in Fig. 3.2(b).

3.2. Modified Spring Analogy for Ultra-Soft Slurry-Like Soil

Terzaghi in 1925 explained the deformation of a saturated soil process by means of a spring analogy. In his analogy, the spring is analogous to the soil mineral skeleton, while water in the cylinder represents the water in the soil voids. Upon application of additional load, the load is carried by water and gradually transferred to the spring. This commencement of load transfer is immediate and the spring takes over the load without delay. The pore pressure dissipation therefore commences immediately.

In the case of the modified spring analogy proposed by Bo et al. in 1997c for slurry-like soil, two separate springs not in a direct contact with each other are used (Fig. 3.3). Therefore load transferred from the water is not

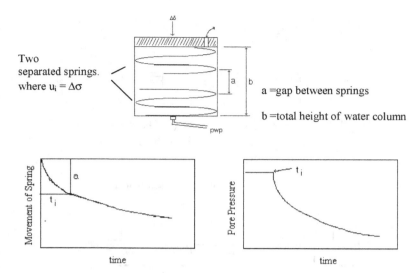

Fig. 3.3. Magnitude and rate of compression.

taken over by the spring until such time when they are in contact. In this case the piston will be moved down without any dissipating of pore pressure until two springs are in contact. This explains the initial deformation behavior of ultra-soft soil in which the soil grains are not touching each other and cannot take over the load from the water. Therefore, this type of soil deforms with no or little effective stress gain. In other words, the deformation with no or little pore pressure dissipation is a possible behavior in slurry-like soil. In this type of slurry-like soil, the early stage of deformation is likely to be controlled by the hydraulic conductivity of soil. In a later stage, when the slurry becomes a soil it would exhibit stress-dependent deformation.

3.3. Magnitude and Rate of Compression

In every compressibility and consolidation study, there are two important approaches: one is how much magnitude would be contributed and another is rate of deformation.

3.3.1. *Magnitude of Compression*

In compression of soil, the magnitude of vertical deformation is directly related to the applied additional load. In the case of vertical compression of ultra-soft slurry-like soil, the magnitude is again directly related to the applied additional load. However, the magnitude can be differentiated into two portions: a viscous and a soil stage.

Therefore, there is a transition point between the slurry stage and soil stage for ultra-soft slurry-like soil. For various high moisture content soils under the same magnitude of load, this transition point from slurry to soil stage may be the same. However, the magnitude of compression in the viscous stage increases with increasing moisture content. For the same moisture content soil under various magnitude of load, transition points may vary due to varying magnitude of loads. The transition point in terms of void ratio is likely to be decreasing with the magnitude of additional load. However, this difference may not be large. Therefore, the magnitude of vertical compression in the viscous stage could be described by the following equation:

$$\Delta H_i = \frac{e_i - e_{t(\Delta\sigma)}}{1 + e_i} \times H_i \qquad (3.1)$$

where ΔH_i = change in thickness in slurry stage; H_i = thickness of ultra-soft slurry-like soil; e_i = initial void ratio of ultra-soft soil; and $e_{t(\Delta\sigma)}$ = transition void ratio which could be a function of additional load if excessive load compared to self-weight is applied.

In one-dimensional consolidation, the magnitude of vertical displacement is computed using parameters from the void ratio versus pressure curve. For ultra-soft soil it is also possible that the void ratio versus pressure relationship can be obtained from one-dimensional compression tests using hydraulic consolidation cell and compression parameters can be obtained. Comparison of possible e-log σ' curves of slurry-like soil and normal soil are shown in Fig. 3.4. The Compression Index in the soil stage is likely to be equivalent to the intrinsic Compression Index of natural clays.

3.3.2. Combined Basic Equation for Viscous Stage and Soil Stage

The total magnitude of settlement comprises a viscous stage (Eq. (3.1)) and a soil stage resulting in a combined equation as follows:

$$\Delta H_{\text{total}} = \frac{e_i - e_t}{1 + e_i} \times H_i + \frac{e_t - e_f}{1 + e_t} H_s \qquad (3.2)$$

where e_f = final void ratio and H_s = thickness of soil at transition point.

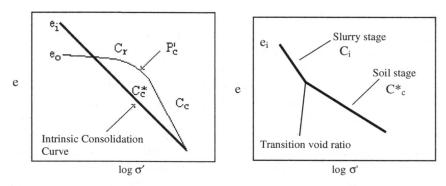

Fig. 3.4. Comparison of e-log σ' curves for natural soil and ultra-soft slurry-like soil.

Therefore, in order to calculate the magnitude of settlement, it is necessary to know the transition void ratio and the final void ratio which are both functions of the additional load. In Terzaghi's simplified equation for elastoplastic soil deformation, only the initial void ratio (natural void ratio e_0) was used. Instead of applying void ratios at the end of the elastic and final stage, he introduced soil compression parameters such as the compression index (C_c), recompression index (C_r) together with an effective stress term. The transition point between the recompression range and compression range was described as a psuedo-preconsolidation pressure which can be determined by a method proposed by Casagrande (1936).

Like Terzaghi's equation the ultra-soft soil compression can be expressed in terms of an ultra-soft soil parameter. Therefore, Eq. (3.1) becomes:

$$\Delta H_i = \frac{C_i}{1+e_i} \times H_i \log \sigma'_t \qquad (3.3)$$

where C_i is the compression index in the ultra-soft soil stage and σ'_t is the effective stress at transition point.

For the soil stage, Terzaghi's one-dimensional equation can be applied without the recompression portion and the initial void ratio for the soil can be taken as the transition void ratio. Therefore, the total settlement of the soil both the ultra-soft and soil stages becomes:

$$\Delta H_{\text{total}} = \frac{C_i}{1+e_i} H_i \log \sigma'_t + \frac{C_c^*}{1+e_t} H_s \log \frac{\sigma'_f}{\sigma'_t} \qquad (3.4)$$

where ΔH_{total} is the total settlement of both stage, C_c^* is the compression index in soil stage and σ'_f is the final effective stress.

It can be seen that in addition to the soil parameters such as the Compression Indices, the transition points in terms of both void ratio and effective stress are required. This will be discussed further in the later chapters.

3.3.3. Time Rate of Compression

The time rate of compression is dependent upon the rate of water flowing out of the soil mass. Since the flow of water follows Darcy's Law, Gibson et al. (1967) used the following equation for saturated sludge:

$$q = n(v_f - v_s) \qquad (3.5)$$

where q = flow rate of water; n = porosity; v_f and v_s = velocities of fluid and solid respectively.

Hence

$$q = -ki \qquad (3.6)$$

where k is the hydraulic conductivity and i is the hydraulic gradient and continuity gives

$$v_s(1-n) + v_f n = 0 \qquad (3.7)$$

Therefore, hydraulic conductivity can be given as

$$k = -\frac{v_s}{i} \qquad (3.8)$$

Been and Sill's (1981) has stated that the average solid velocity for an element of soil is simply the change in height of the soil element. As such the settlement during the slurry stage can be simply worked out from the average hydraulic conductivity. The hydraulic head can be taken as an additional load since there will not be any significant pore pressure change during slurry deformation; since velocity is changing with time, k is also changing with time. Hence

$$-\nu_s = ki \qquad (3.9)$$

$$-\frac{\Delta H}{t} = ki \qquad (3.10)$$

This type of compression would end at the time the slurry changes to soil. After this point the rate of settlement can be worked out using Terzaghi's conventional theory. Therefore, in order to work out time rate of compression in the slurry stage, the hydraulic conductivity of slurry is required. However, it should be noted that the hydraulic conductivity of ultra-soft soil may change during compression.

3.3.4. *Slurry to Soil*

To calculate the total compression of ultra-soft soil and time rate of settlement in the different stages, the transition point is necessary to find out. When transition point is available, the magnitude of settlement in two different stages such as slurry stage and soil stage could be worked out using two basic equations. This transition point will also suggest the commencement of significant pore pressure dissipation. To determine the compression behavior of ultra-soft soil, various experimental tests could be carried out. The possible types of experimental tests are as follows:

(1) One-dimensional compression tests with large diameter consolidometer equipped with pore pressure transducers.
(2) One-dimensional compression tests with small scale consolidometer.
(3) One-dimensional compression tests with Rowe cell hydraulic consolidometer.
(4) Constant rate of loading tests.
(5) Constant rate of strain tests.

From the above tests the compression parameters in the slurry stage such as the transition points in terms of void ratio and stress, hydraulic conductivity in the slurry stage, hydraulic conductivity change index, and compression index in the slurry stage can be obtained.

3.3.5. *Coordinate System and Governing Equation*

3.3.5.1. *Coordinate System*

To calculate the magnitude and rate of compression, a coordinate system and governing equations must first be selected.

There are four coordinates system previously used as follows:

(i) Eulerian (Morgenstern and Nixon, 1971).
(ii) Convective (Gibson *et al.*, 1967).

(iii) Lagrangian (Townsend and McVay, 1990; Monte and Krizek, 1976).
(iv) Reduced (Gibson et al., 1981).

Among these, the Eulerian coordinate is commonly used in small strain consolidation theories since it defined a point in space with respect to an origin fixed in space. In large strain the coordinates of a boundary will change during the process of compression at a rate that is solely determined by the process itself (Lee and Sills, 1981). In Lagrangian coordinates a point is labeled in terms of the amount of solid material between that point and an origin fixed in the material. Therefore, the coordinates move with the material so that a boundary always has the same coordinates. The Eulerian coordinate can be related through the void ratio as follows:

$$X = \int_0^Z (1+e)dz \qquad (3.11)$$

However, for large strains without self-weight effects Gibson et al. (1967) used Lagrangian coordinates while Lee and Sills (1981) solved a similar problem using Eulerian coordinates with a moving boundary.

Lee and Sills (1981) has stated that the effect of self-weight is less marked when an external load is applied to the soil layer. In our approach, the applied external load to the ultra-soft soil was much greater than the self-weight. Therefore, the contribution from the self-weight consolidation term was considered to be negligible. The Lagrangian coordinate system was also chosen for the analyses.

3.3.5.2. Governing Equation

The governing equation adopted is based on three basic laws of mass conservation, balance of momentum, and energy conservation. Since heat phenomenon can be ignored in compression problems, the two laws such as mass conservation and balance of momentum shall be sufficient to explain the compression magnitude and rate. To formulate these laws into a mathematical expression, the definition of a material point is required referring to a coordinate system to be used. Since we are dealing with large strain compression, the selected coordinate system should allow for parameter and boundary changes during the compression such as change of hydraulic conductivity, void ratio, and movement of element and boundary. As explained by Gibson et al. (1967), distinguishing two coordinate systems is important in finite large strain compression. Therefore, Gibson et al. (1967) have differentiated two coordinate systems with Fig. (3.5).

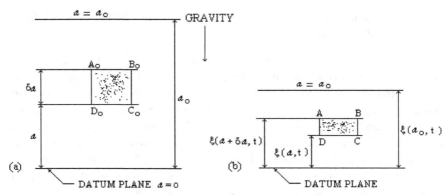

Fig. 3.5. Lagrangian and convective coordinates: (a) initial configuration at $t = 0$, (b) configuration at time "t" (after Gibson et al., 1967).

In which element of soil ABCD at $t = 0$ is explained with independent variable "a" and same element at some subsequent time "t" shall be located at unknown distance described by dependent variable. However, using Lagragian coordinates, exact location $\xi(a_{0,t})$ could be ignored and boundary can always be identified ($a = a_0$).

By assuming soil grains being homogeneous and both the pore fluid and solids are incompressible, the equation governing the rate of change in void ratio is as follows (Gibson et al., 1967):

$$\pm \left(\frac{\rho_s}{\rho_f} - 1\right) \frac{d}{de} \left(\frac{k(e)}{1+e}\right) \frac{\delta e}{\delta z} + \frac{\delta}{\delta z} \left(\frac{k(e)}{\rho_f(1+e)} \frac{d\sigma'}{\delta e} \frac{\delta e}{\delta z}\right) + \frac{\delta e}{\delta t} = 0 \quad (3.12)$$

Therefore, if the density of solid, fluid, and hydraulic conductivity changes with the void ratio and the void ratio changes with position of element (δz) is known, rate of void ratio change can be calculated. This large strain equation would be applied in our study and whether the effective stress term can be neglected in the viscous stage would be discussed.

3.3.6. Numerical Approach

To solve the governing equation numerically, finite difference equations in implicit form were used. Firstly, the variation of hydraulic conductivity has to be determined from laboratory test. Lee and Sills (1981) proposed the following relationship:

$$\frac{k}{\rho_f} = K_0(1+e) \qquad (3.13)$$

where K_0 is constant.

However, Gibson et al. (1967) defined the coefficient of consolidation for large strain as

$$C_F = \frac{k}{\rho_f(1+e)} \frac{d\sigma'}{de} \qquad (3.14)$$

This parameter takes into account the stress–strain relationship and hydraulic conductivity. Lee and Sills (1981) reported that this parameter C_F is less variable than any other individual parameter in the equation. Therefore, the equation for large strain becomes

$$\frac{de}{dt} = C_F \frac{d^2e}{dz^2} \qquad (3.15)$$

This simplified equation can be written in implicit finite difference equation as follows:

$$\frac{e_{i,t+\Delta t} - e_{i,t}}{\Delta t} = C_F \frac{\left[\frac{e_{i-1} - 2e_i + e_{i+1}}{(\Delta z)^2}\right]_t + \left[\frac{e_{i-1} - 2e_i + e_{i+1}}{(\Delta z)^2}\right]_{t+\Delta t}}{2} \qquad (3.16)$$

$$e_{i,t+\Delta t} = \frac{(e_{i-1} + e_{i+1})_{t+\Delta t} + e_{i,t}}{2\left(1 + \frac{1}{\beta}\right)} \qquad (3.17)$$

where

$$e_{i,t} = \left[e_{i-1} - 2\left(1 - \frac{1}{\beta}\right)e_i + e_{i+1}\right]_t \qquad (3.18)$$

$$\beta = \frac{C_F \Delta t}{(\Delta z)^2} \qquad (3.19)$$

The above implicit finite difference equations will be used for the analyses in this study.

Chapter 4

CHARACTERIZATION OF PHYSICAL PROPERTIES AND MINERALOGY OF THE SOIL

4.1. Introduction

To improve the understanding of ultra-soil compressibility, author has carried out comprehensive experimental program, which consists the types of tests suggested in the later chapters. Findings from these will be explained in the subsequent few chapters. In addition, findings from these experimental results will be used to formulate the necessary equations and suitable theory of compressibility that can be applied to ultra-soft soil compression. In the experimental program, ultra-soft soils were prepared from the samples collected from the slurry pond. The characteristic of such soils used in the experimental programme is described in details in the following paragraphs.

The soil being studied is an ultra-soft soil from a deep deposit of silty clay slurry at a major reclamation project in Changi East, Singapore. It has a high water content greater than its liquid limit. The strength of the top half of deposit is extremely low and may not have measurable effective stress in the *in situ* condition. This soil may be undergoing self-weight consolidation. A tall column of this soil if left undisturbed will probably develop a structure during the self-weight consolidation process. However, it would take a long period of time for the soil to possess an effective stress that is equivalent to its effective overburden pressure. When it is subjected to either a high pressure or hydraulic head gradient, the soil will probably flow and behave like a viscous liquid. Samples for the laboratory tests were taken from that slurry pond in which slurries were submerged under 7 m depth of water. The predominant mineral is kaolinite. It is deposited in the slurry pond since 1970s when the sand quarry wash from Tampines sand mine in Singapore was disposed hydraulically there.

4.2. Physical Properties

The materials used for testing consist of 1% sand, 48% silt, and 51% clay. The mean grain size D_{50} is 0.002 mm. The typical grain size distribution interpreted from hydrometer tests is shown in Fig. 4.1. The liquid limit determined using fall cone test is between 72% and 108% and plastic limit is between 26% and 28%. The corresponding plastic index varies between 46% and 80%. The specific gravity is about 2.68. When the data are plotted on the plasticity chart, they lie above the "A" line and can be classified as a high plasticity clay (Fig. 4.2). The activity of this soil is about 0.51–0.55 and can be considered as an inactive clay.

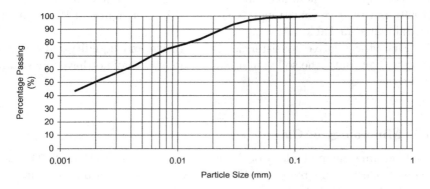

Fig. 4.1. Typical grain size distribution of tested sample.

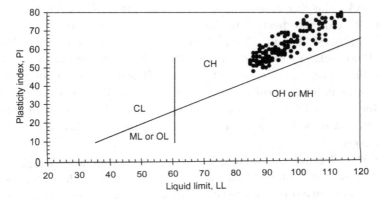

Fig. 4.2. Physical characteristics of study soil on characteristics chart.

Table 4.1. Relative consistency (Cr) and liquidity index (LI) of tested soils.

Moisture content (%)	LL	PL	PI	Relative consistency (Cr)			Liquidity index (LI)		
				(a)	(b)	(c)	(a)	(b)	(c)
130	—	—	—	−1.26	−0.79	−0.28	2.26	1.79	1.28
150	(a) 72	26	46	−1.70	−1.14	−0.53	2.70	2.14	1.58
170	(b) 85	28	57	−2.13	−1.49	−0.78	3.13	2.49	1.78
190	(c) 108	28	80	−2.56	−1.84	−1.03	3.56	2.84	2.03

Note: a, b, c refer to three different liquid limits.

In this study, samples were prepared with four different water contents. The corresponding liquidity index (LI) defined by Lambe and Whitman (1969) and relative consistency (Cr) defined by Terzaghi and Peck (1948) are tabulated in Table 4.1.

Since the relative consistencies are negative and liquidity indices are greater than unity, the soil is in a liquid state and may not have any shear strength.

4.3. Mineralogy of Clay

To determine the mineralogy of the soil, both X-ray diffraction (XRD) and scanning electron microscopy (SEM) tests were carried out on the collected samples by the British Geological Survey, UK.

The XRD analyses were carried out using a Philips PW1700 series diffractometer equipped with a cobalt-target tube and operating at 45 kV and 40 mA. The clay mineralogies were determined by scanning air-dried, glycol-solvated and heated to 550°C/2 oriented mounts from 1.5° to 32° 2θ at 0.48° 2θ/min. Diffraction data were analyzed using Philips APD 1700 software coupled to a JCPOS database running on a DEC Micro Vax 2000 micro-computer system. Based on XRD analysis the major contents of mineral were found to be kaolinite and smectite. Mica and chlorite were found as minor minerals. The XRD results of samples from 2-m, 5-m and 8-m depth are shown in Figs. 4.3, 4.4 and 4.5, respectively.

In the electron microscope scanning, specimens were examined in a Cambridge instrument SEM S250 MK 1 fitted with a link 860 A energy dispersive X-ray analysis (EDXA) system which provided qualitative chemical information from the areas of interest.

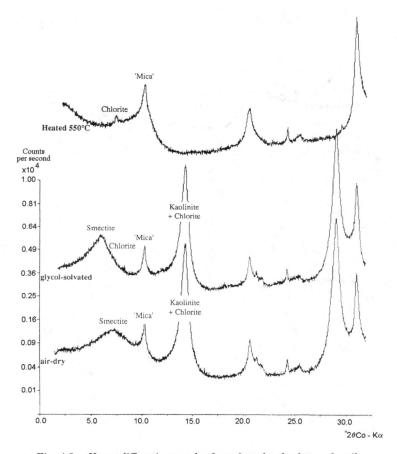

Fig. 4.3. X-ray diffraction results from 2-m depth ultra-soft soil.

Results showed that the samples normally consisted of ragged clay flakes, which were poorly consolidated and had significant high porosity. Individual clay flakes were found to be 10 μm across. In addition to the flakes, some discrete kaolinite subspherical pallets of up to 10 μm long were observed. EDXA indicated that the kaolinite was intimately mixed with iron oxides. Authigenic Pyrite was present as isolated, sub-rounded grains and was typically associated with micron-scale, predominantly structureless to poorly developed flakes of clay with a Fe, K, Mg, Al, Si chemistry. The SEM photomicrographs are shown in Fig. 4.6. It can be seen in the photo that the specimen was highly porous.

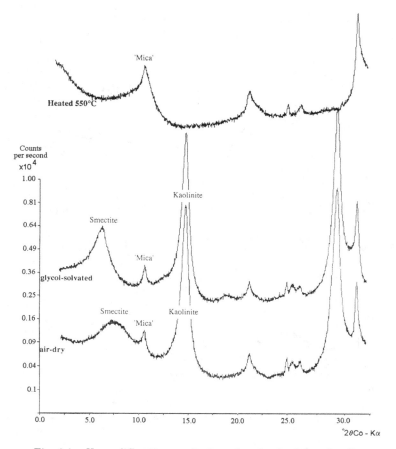

Fig. 4.4. X-ray diffraction results from 5-m depth ultra-soft soil.

4.4. Chemistry of Pore Water

Pore water samples were collected from the slurry pond at the foreshore area in a marine environment, using the BAT Groundwater sampler. The pore water was found to be in a slime condition. The total dissolved solids were between 18,000 and 21,000 ppm. The major cation was sodium (Na^+) and the major anion was chloride (Cl^-). The pH value was around 8, which indicated an alkaline condition. Potassium (K^+), magnesium (Mg^+), and calcium (Ca^{++}) were found as the minor cations. Sulfate (SO_4^{--}) was found as the minor anion. No trace of nitrate was found. The electric conductivity

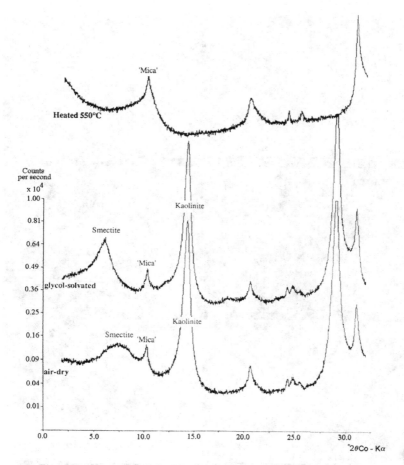

Fig. 4.5. X-ray diffraction results from 8-m depth ultra-soft soil.

of the pore water was found to be about 30 ms/cm. Details of some of the chemical content of the pore water are shown in Table 4.2.

4.5. Preparation of Ultra-Soft Sample

Samples to be tested were grabbed from the man-made slurry pond, which is located in the foreshore of eastern Singapore. Since the slurry consisted of fine materials at 5–7 m depth contained in a bunded area and submerged under seawater, the samples collected were probably in a saturated state at the time of collection. The collected samples were stored in a large

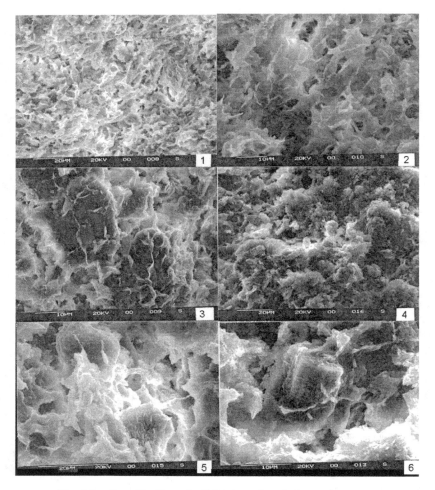

Fig. 4.6. Magnification of soil fabric under electron microscope.

diameter plastic tank and kept inside the laboratory sample storage room at a constant temperature of 27°C.

For each test, samples about three times of required quantity were obtained from the storage tank. Samples were mixed with diluted seawater, which has more or less the same total dissolved solid concentration as the natural pore water. In order to obtain the exact moisture content in the ultra-soft slurry-like soil, firstly the natural moisture content in the slurry-like soil was checked. In the moisture content tests, the correction for salt

Table 4.2. Chemistry of pore water from tested samples.

Parameter	No. 1 Sample	No. 2 Sample
TDS (ppm)	18,510	21,305
Na (ppm)	6,587	6,254
K (ppm)	368.2	664.8
Mg (ppm)	655.9	895.5
Ca (ppm)	301.7	82.29
Cl (ppm)	10,575	11,495
SO$_4$ (ppm)	1,626	1,950
EC (ms/cm)	29.6	31.7
pH	7.86	7.98

content were made applying the following equation suggested by Imai *et al.* (1979):

$$W = \frac{1+\beta}{1-\beta W_m} \times W_m \qquad (4.1)$$

where W is the corrected moisture content, W_m is the measured water content, $\beta = \frac{W_c}{W_w}$ where W_c is the weight of salt in the water and W_w is the weight of water.

The salt correction is necessary to obtain correct moisture content and void ratio. If salt correction is not done the underestimated moisture content and void ratio would be obtained. The significant difference of the moisture content values with and without salt corrections is shown in Fig. 4.7.

Void ratios were calculated from the corrected moisture content. To obtain a soil with the exact required moisture content, a known weight of

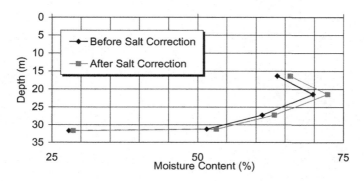

Fig. 4.7. Comparison of moisture content with and without salt corrections.

diluted seawater is added using the following equation:

$$M_{WR} = \frac{M_s}{W_{W1} + 1}(W_{W2} - W_{W1}) \qquad (4.2)$$

where M_s = total weight of soil; W_{W1} = current water content; W_{W2} = required water content; and M_{WR} = required quantity of water by weight to be added to obtain the required Moisture content.

Samples were directly prepared in the cell used for the testing to obtain the required water content. Thorough mixing and stirring were carried out to remove any air trapped in the sample.

Chapter 5

COMPRESSION TESTS WITH LARGE SCALE CONSOLIDOMETER

In this study, two types of compression tests using a large consolidometer equipped with linear vertical displacement transducers (LVDT), pore pressure transducers, and total earth pressure cells were carried out. With these instruments, the compression behavior of ultra-soft soil as well as pore pressure responses were monitored throughout the tests. The first type used two-step high pressure loading, 100 and 190 kPa. The second type used incremental step loading starting from an applied pressure of 12 to 44 kPa.

The high-pressure immediate loading simulates the field condition where filling is placed within a short time. The incremental step loading simulates the slow stage filling.

The settlement, pore pressure, and lateral displacement of the soil sample were investigated in the study. Post-mortem investigations were carried out on the tested samples.

5.1. Method of Testing

The compression tests were carried out using two different methods of loading:

(i) Two-step high pressure loading.
(ii) Step incremental loading.

The aim of the two-step high pressure loading is to study the deformation and creep behavior of the ultra-soft slurry-like soil, which is confined in the cell. This condition is similar to the case of reclamation over a slurry pond or waste pond in which instantaneous loading is applied on the soft slurry. A case study of such a reclamation has been reported by Bo et al. (1998). The pressures applied in two-step loading were 100 and 190 kPa. The incremental step loading was used to simulate stage filling in the field. The load increments were 12, 19, and 44 kPa.

Pore pressures were measured at several locations at different distances from the drainage boundaries. Average settlements were measured at the top of the sample. During testing, the total pressures were measured by the earth pressure cells, installed at the top only for the two-step high pressure loading and at the top and bottom of the sample for the incremental step loading. With this measurement, the wall friction around the soil sample can be determined. The friction between the cylindrical wall and loading plate was checked before carrying out the tests and was found to be about 10 kPa. In the incremental step loading test, the rate of loading was determined based on the rate of pore pressure dissipation. For the two-step high pressure loading tests, each test was carried out until the primary consolidation was essentially completed based on the pore pressure measurement.

At the end of the test the index property, strength, and consolidation tests were carried out on samples cut from the consolidated soil mass at various distances from the central drain. The results were compared and discussed.

5.2. Description of Apparatus and Sample Preparation

The apparatus is a large diameter consolidation cell with an inside diameter of 495 mm and a height of 1000 mm. The cell is made of stainless steel with smooth inside wall. The wall was also coated with silicon grease before placing the sample. This is an upgraded version of the large cell reported by Bo et al. (1999). A total earth pressure cell (vibrating wire type) was placed on the base plate with the sensitive side facing upward. Two pore pressure transducers, WF 17060 manufactured by Wykeham Farrance, were installed on the sidewall, 220 and 420 mm above the bottom of the cell. The accuracy of measurement is 1 kPa.

The prefabricated vertical drain (Colbond CX 1000), having a width of 100 mm and a thickness of 5.3 mm was placed vertically in the cell and anchored at the bottom plate. The filter of the vertical drain has a hydraulic conductivity of 11×10^{-4} m/s while the core has a discharge capacity of

$70 \times 10^{-6}\,\mathrm{m^3/s}$ in straight configuration and $20 \times 10^{-6}\,\mathrm{m^3/s}$ in 25% buckled condition at a confining pressure of 100 kPa. The filter has an apparent opening size O_{95} (size of opening in a geotexile for which 95% of opening are smaller) smaller than 75 μm. The top end of the vertical drain was connected to a drainage pipe so that the quantity of water flow from the drain can be measured with a measuring cylinder. A similar configuration of apparatus was used by Kim et al. (1995) in their study on the consolidation of dredged clay improved by horizontal drains. The slurry taken from the slurry pond (silt pond) was mixed thoroughly with seawater to the desired water content and poured into the consolidation cell. For the two-step high pressure loading tests the sample height was 750 mm. For the incremental step loading test the sample height was 700 mm. The slurry was stirred to remove the air trapped in the slurry.

The basic physical properties of the tested samples are shown in Table 5.1. Four numbers of miniature pore water pressure transducer PDCR 81 manufactured by Druck Ltd. were pushed into the soil at different distances such as 100 and 200 mm from the vertical drain. The accuracy of measurement of these transducers is 1 kPa.

Two transducers were installed at 200 and 400 mm above the bottom of the cell for the two-step high pressure loading test and 240 and 560 mm for the incremental step loading test. The details of the apparatus and transducer locations before the commencement of the test are shown in Fig. 5.1. The application of a load to the soil was carried out by applying compressed air pressure to a rigid diaphragm which was installed on top of the tested sample.

Table 5.1. Basic physical properties of tested soils.

Parameter	2-step high pressure loading test	Step incremental loading test
Moisture content (%)	132	175
Bulk density (KN/m^3)	13.33	12.85
Dry density (KN/m^3)	0.59	0.48
Specific gravity	2.67	2.69
Void ratio	3.55	4.65
Clay content (%)	60	50
Organic content (%)	0.90	0.70
Liquid limit (%)	73	85
Plastic limit (%)	27	28

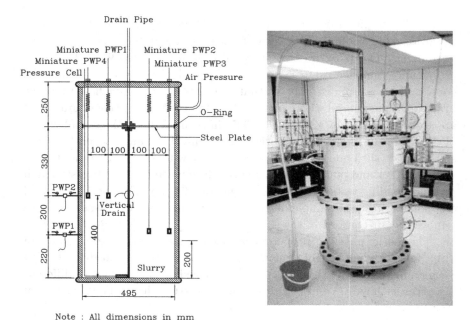

Note : All dimensions in mm

Fig. 5.1. Large diameter consolidometer equipped with pore pressure transducers and total pressure cell.

5.3. Two-Step High Pressure Loading Tests

5.3.1. *Deformation Behavior During First Step Loading*

The first step loading of 110 kPa was applied to the ultra-soft soil. No backpressure was applied in the test. Pore pressures were registered immediately in all transducers varying between 102.6 and 109.4 kPa. The measured pore pressure corresponded to the applied pressure of 110 kPa and the hydrostatic pressure. The adjustment was made for friction of 10 kPa. Therefore, different transducers measured slightly different pore pressure values due to the difference in locations. However, the miniature PWP transducers 1 and 4 measured 5–6 kPa higher than expected. The total pressure cell measured the pressure due to the total soil weight plus the given pressure. The measured value was approximately equal to the expected value. A summary of initial measurement compared with the expected values are shown in Table 5.2.

Table 5.2. Summary of measurements in two-step high pressure loading.

	First loading (100 kPa)			Second loading (190 kPa)			Gain in effective stress[a] (kPa)
	Expected reading (kPa)	Initial reading (kPa)	Final reading (kPa)	Expected reading (kPa)	Initial reading (kPa)	Final reading (kPa)	
Mini-PWP 1	103	108.1	12.2	102.2	135.5	52.7	137.3
Mini-PWP 2	105	105.2	10.6	100.6	130.5	63	127
Mini-PWP 3	105	106.4	12.9	102.9	133.3	67.4	123
Mini-PWP 4	103	109.4	6.8	96.8	137.8	74	116
PWP 1	105	105.2	—	—	—	—	—
PWP 2	103	102.6	—	—	—	—	—
Pressure cell	110	111.2	42	200	185.7	160.7	—
Air pressure	110	110	110	200	200	200	—

[a](Total pore pressure) without considering applied pressure losses due to friction.

5.3.2. Pore Pressure and Settlement

Pore pressure measurements from all six transducers were studied carefully. Four of them (mini-PWP 3 and 4, were 200 mm away from the vertical drain and PWP 1 and 2 are on the wall and about 250 mm away from the vertical drain) measured no or little pore pressure dissipation in the first 10 days of measurement (Fig. 5.2). The settlement after 10 days was about 150 mm which is equal to 20% strain (Fig. 5.3). This phenomenon was consistent with the earlier finding of little or no effective stress gain in the early stage of slurry deformation measured in a similar type of test (Bo et al., 1999). However, the remaining two transducers mini-PWP 1 and 2 located at 100 mm away from the vertical drain responded differently. Mini-PWP 1 started measuring pore pressure dissipation within about 12 h whereas the pore pressure dissipation at mini-PWP 2 commenced on the sixth day of measurement (Figs. 5.2(e) and 5.2(f)). The settlement at 12 h and 6 days were 22 and 120 mm which were about 3% and 16% of strains, respectively. The difference in pore pressure dissipation from the two mini-PWP which were initially located at the same distances away from the vertical drain could be due to the lateral movement of the soil during the early stage of deformation. This caused the mini-PWP 1 to be closer to the vertical drain than the mini-PWP 2.

Leroueil et al. (1986) reported the differences between $(e - \sigma')$ paths of soil sublayers located at different distances from the drainage boundary. Hight et al. (1987) explained that these variations are due to strain rate effects since the strain rate in each layer is varying. Mesri et al. (1995)

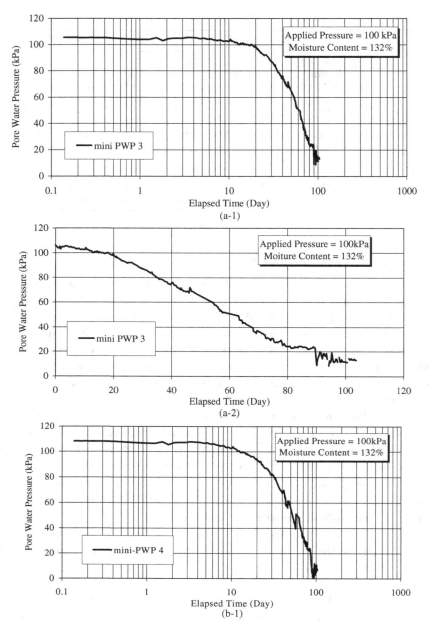

Fig. 5.2. Pore pressure measurement during the first step high pressure loading: (a) mini-PWP 3; (b) mini-PWP 4; (c) PWP 1; (d) PWP 2; (e) mini-PWP 1; (f) mini-PWP 2. (1 and 2 refer to semi-log scale and arithmetic scale respectively.)

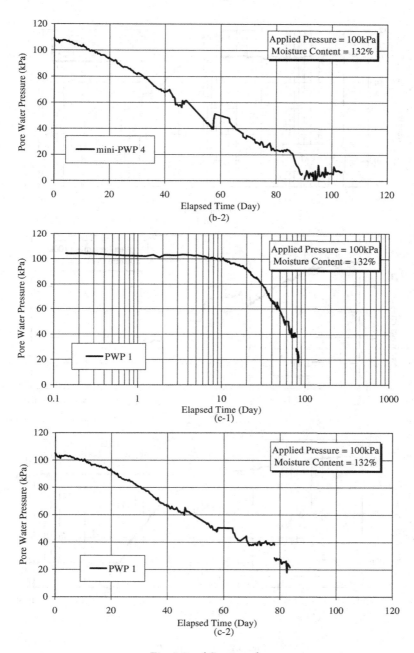

Fig. 5.2. (*Continued*).

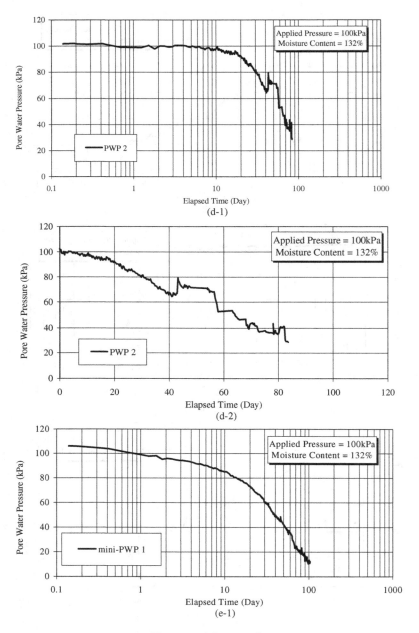

Fig. 5.2. (*Continued*).

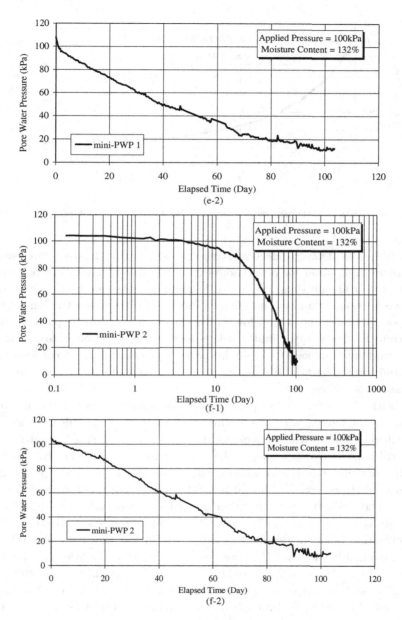

Fig. 5.2. (*Continued*).

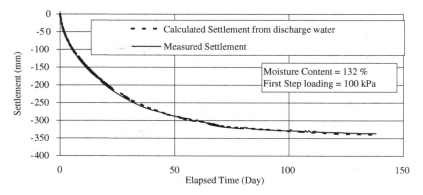

Fig. 5.3. Vertical settlement measurement during the first step high pressure loading.

describes the delayed effective stress gains, i.e. the delayed pore water pressure dissipation for a few hours from the subsoil layers located away from the drainage boundary for a 500 mm drainage path. However, the above phenomenon alone may not be able to fully explain the long delay of pore pressure dissipation for the relatively short drainage path of 100–200 mm in this study. Similar delays in the consolidation of slurry were observed by Tanaka (1997) in his study on nonuniformly consolidated ground around vertical drains. However, no explanation was given for the cause of the delay.

Although there was no pore pressure dissipation, the volume of water drained out agreed with the volume change of the slurry determined by the displacement measurement, as shown in Fig. 5.3. This indicates that the volume change occurring in the slurry is due solely to the reduction of water, but this reduction of water does not lead to a reduction of pore water pressure in the slurry. The initial vertical displacement rate is comparatively high. The drained creep (the term defined by Been and Sills (1981)) was explained by Mesri's (1987) universal law of compression. However, the deformation behavior detected in this study in the initial stage could not be explained by this law. It is possible that this rate of settlement, i.e. the rate of water draining out, is only related to the hydraulic conductivity of the filter and soil surrounding the drains and the hydraulic gradient induced by applied pressure. A similar behavior was reported by Sills (1995) on the self-weight consolidation of slurry. In 1968, de Josselin de Jong noted that decrease in volume must be due to the expulsion of water possibly within the flocculated structure of the soil sediment.

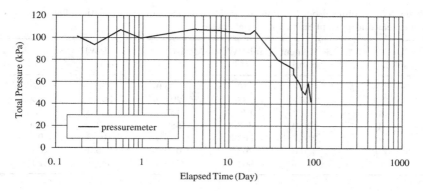

Fig. 5.4. Total pressure measurements during the first step high pressure loading.

The pressures measured from the total earth pressure cell were essentially constant until 20 days after loading. At that time slurry-like soil has been compressed by about 200 mm (i.e. 27% strain) and the remaining thickness of the soil was only about 550 mm. After 20 days the total pressure measured at the bottom of the sample started to decrease and after 90 days, the measurement was only about 60 kPa as shown in Fig. 5.4 and Table 5.2. This could be due to an increase in the wall friction along the sample after the sample became soil-like. During the early stages, the wall friction was negligible. From 20 days to the end of the first step loading, about 139 mm of settlement had occurred.

5.3.3. Comparison of Degree of Consolidation

The degrees of consolidation were computed from both settlement and pore pressure measurements, including the settlement that occurred during the "no or little dissipation" period. A much faster rate of consolidation was noted from the settlement measurement. At a given time, the degree of consolidation computed from the settlement was much higher than that computed from the pore pressure measurements. There is a discrepancy between rates of settlement and pore pressure decay for natural clay. The pore pressure usually lags behind the surface settlement. This behavior was also reported by Hansbo et al. and Jamiolkowski (1981), Leroueil (1986b), Hight et al. (1987), Mikasa (1995), and Jardine and Hight (1987). However, the difference in the degree of consolidation based on settlement

Table 5.3. Comparison of time at various degrees of consolidation based on the settlement and pore pressure measurements.

Degree of consolidation (%)	Settlement (mm)	Mini-PWP 1 (h)	Mini-PWP 2 (h)	Mini-PWP 3 (h)	Mini-PWP 4 (h)
5	8.3	10.0	40.0	296.7	176.7
10	21.7	36.7	200.0	503.3	330.0
20	63.3	170.0	487.0	750.0	563.3
30	130.0	363.3	682.0	923.3	936.7
40	216.7	563.3	835.0	1,103.3	936.7
50	330.0	763.3	1,020.0	1,315.3	1,050.0
60	476.7	963.3	1,272.0	1,411.1	1,295.6

and pore pressure in the test was much greater than that normally found in natural clays.

Among the mini-PWPs, mini-PWP 1 again measured a faster rate of dissipation than the others while mini-PWP 3 and 4 which are located away from the vertical drain, had a slower rate of dissipation. Mini-PWP 4 measured slightly faster dissipation than mini-PWP 3.

Table 5.3 shows that the degree of consolidation measured at various soil elements is only 5% as measured by the pore pressure transducers. It is obvious that the soil elements away from the drain took longer than those close to the drain.

5.3.4. Deformation Behavior During Second Step Loading

When all the pore pressure transducers registered dissipation of more than 90% of initial excess pore pressure, the second stage loading of 90 kPa was applied. Therefore, the total applied pressure was 200 kPa. The total pressure registered at the bottom of the sample immediately after loading was 185.7 kPa. Therefore, there was a reduction of about 15 kPa due to friction. Although the increment of loading was only 90 kPa, the increase in pore pressure and total pressure at the bottom of the sample were about 120 kPa. As explained in the earlier section, the reduction of load to about 30 kPa occurred in the first step loading test due to the build up of wall friction. When the load increased from 110 to 200 kPa, the wall friction broke. The measured pore pressure increased by 120 kPa (90 kPa + 30 kPa) instead of 90 kPa. The difference of about 30 kPa seemed to be from the first step of loading.

The second stage was carried out only up to 23 days due to the problem in pore pressure measurement. After that time the pore pressures were either maintained at the level or fluctuated throughout the measurement with marginal dissipation of pore pressure (Fig. 5.5).

Although the pore pressure at mini-PWP 3 and 4 remained relatively constant after 7 days and mini-PWP 1 and 2 after 1 day, the settlement continued until 15 days.

Some gases were detected coming out from the drainage pipe. The pore pressure increased rapidly as can be seen in Figs. 5.5(a)–5.5(d). The fluctuation of the measured pore pressure was likely due to gas development and cracking of soil. Up to 13 mm of settlement was measured on day 17. About 0.5 mm of swelling was measured after day 17 (Fig. 5.6).

The settlement that occurred in the second stage was much smaller than the settlement in the first stage. Most of the settlement that occurred in the first stage could be attributed to the compression of the clay slurry in the slurry stage.

5.3.5. Post-Mortem Investigation on the Compressed Sample

At the end of the test, the clay block was extruded and cut open to view the condition of the vertical drain and the transducer locations. The vertical drain was deformed as shown in Fig. 5.7 and mini-PWP 1 and 4 were displaced toward the drain. Both mini-PWP 1 and 4 moved toward the drain by about 95 mm. They were located only 230 mm away from the bottom compared to 400 mm initially. Mini-PWP 2 and 3 moved about 35 mm toward the drain. This lateral movement is likely to be caused by the flow of ultra-soft soil during slurry deformation. They were located only 120 mm above the bottom as compared to 200 mm initially.

It appeared that the bottom 200 mm of the sample settled 80 mm. The next 200 mm layer settled 90 mm, and the top 350 mm layer settled 178 mm. It can be concluded that more than half of the settlement came from the upper half of the specimen. The strains of these layers are summarized in Table 5.4. The highest strain occurred from the top 350 mm layer. In addition to the vertical displacement, lateral migration of soil toward the drain was observed. This finding is consistent with that reported by Tanaka (1997) on soil migration during consolidation.

However, the void ratios calculated from the settlement of the middle and bottom layers are higher than the measured void ratios, whereas the

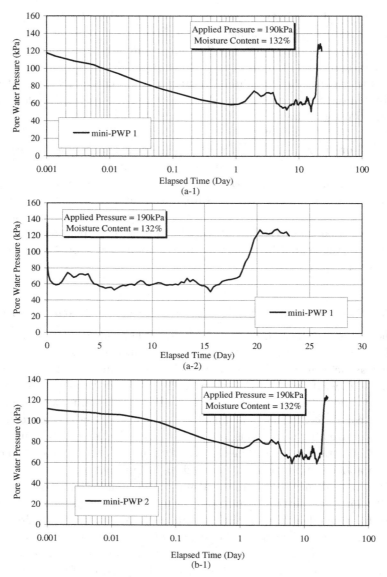

Fig. 5.5. Pore pressure measurement during the second step high pressure loading: (a) mini-PWP 1; (b) mini-PWP 2; (c) mini-PWP 3; (d) mini-PWP 4. (1 and 2 refer to semi-log scale and arithmetic scale, respectively.)

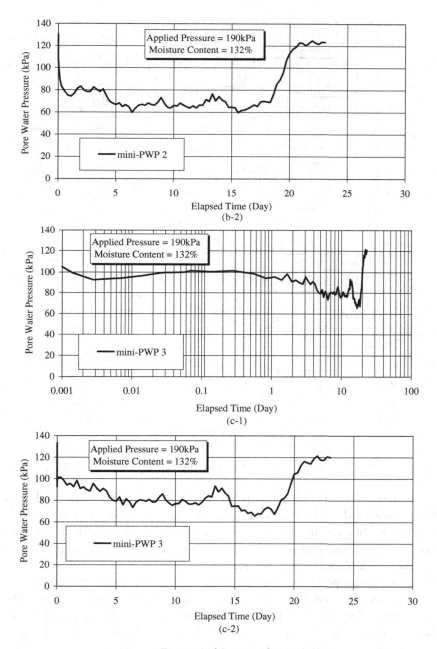

Fig. 5.5. (Continued).

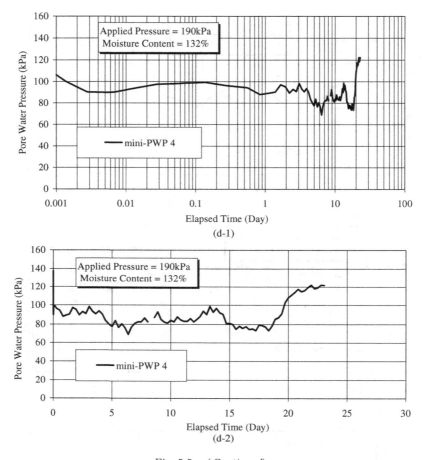

Fig. 5.5. (*Continued*).

calculated void ratio at the top layer is lower than the measured void ratio. Measured void ratios after compression are shown in Fig. 5.11. It is possible that the middle and bottom layer settlements calculated from the piezometer movements are smaller than the actual movements due to the lateral displacements of the piezometers.

Physical properties, strength, and consolidation tests were conducted on samples obtained from different locations of the clay block after the compression test. Figure 5.10 shows the content of clay, silt, and sand. It can be seen that higher clay content was found near the drainage. The migration of fines toward the drain was evident. This could be due to the high gradient

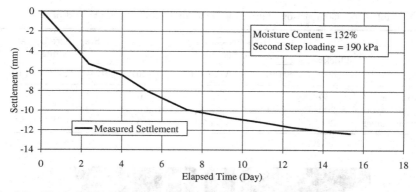

Fig. 5.6. Vertical settlement measurement during the second step high pressure loading.

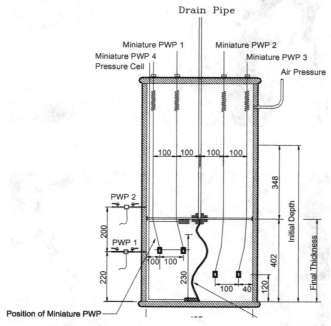

Fig. 5.7. Shape of PVD and location of mini-PWP after the completion of the compression test.

Table 5.4. Comparison of strain from subdivided layers at the end of tests.

	Thickness (mm)	High Pressure two-step loading		Thickness (mm)	Step incremental loading	
		Strain (%)	Void ratio		Strain (%)	Void ratio
Top layer	350	51	1.236	140	51	1.77
Middle layer	200	45	1.5	320	44	2.16
Bottom layer	200	40	1.73	240	49	1.85

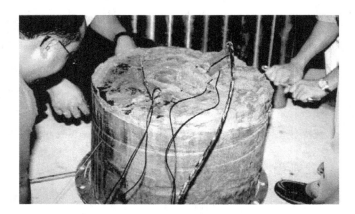

Fig. 5.8. Closed up view of the compressed sample.

Fig. 5.9. Cutting of samples from the compressed block sample.

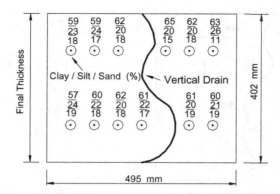

Fig. 5.10. Grain size distribution of samples after compression.

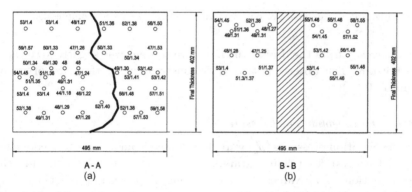

Fig. 5.11. Measured moisture contents and void ratios after compression. (53/1.4 = Moisture content/Void ratio.)

of pore pressure. Locations of samples used in the consolidation tests and other physical tests are shown in two Secs. A-A and B-B in Figs. 5.11 and 5.12. Section A-A is oriented perpendicular to the drain, whereas Sec. B-B is parallel to the drain. The sample orientations in the oedometer test are shown in Fig. 5.12.

5.3.5.1. *Moisture Content and Bulk Density*

The moisture content after compression varied between 45.59% and 59.16%. Therefore, the void ratio varied between 1.22 and 1.59 (Fig. 5.11). A significant variation of void ratios was noted. The lowest moisture content was found near the drain. The moisture content increased with increasing

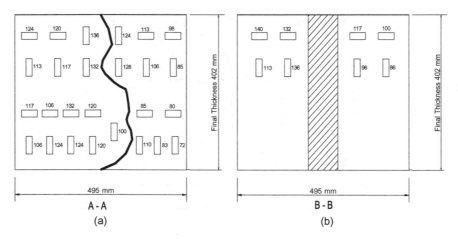

Fig. 5.12. Measured preconsolidation pressure in kPa after compression.

distance away from the drain. The moisture content was slightly lower near the bottom. The observed pattern is consistent with those reported by Tanaka (1997).

5.3.5.2. Preconsolidation Pressure

Preconsolidation pressures (p'_c) were determined from laboratory oedometer tests. Casagrande (1936) graphical method was adopted in the determination of p'_c values. The variations of p'_c within the clay block are shown in Fig. 5.12. In general preconsolidation pressure is higher near the drain and lower away from the drain (Fig. 5.12). The preconsolidation pressures measured were lower than the gain in effective stress calculated from pore pressure dissipation without taking into consideration the load reduction due to wall friction in the first step loading. The total applied load was 200 kPa but the effective stress increased by only 80–136 kPa. It was interesting to see that the effective stresses were lower at the bottom part as compared to those from the top part. This is probably due to the development of wall friction. The effective stress gain of soil elements at different locations are shown in Table 5.5.

5.3.5.3. Undrained Shear Strength

Undrained shear strengths were measured using the laboratory vane and the Swedish fall cone. The results are shown in Fig. 5.13. The strength is

Table 5.5. Effective stress gain of soil element at different locations.

	Top layer		Middle layer		Bottom layer	
	Left of drain	Right of drain	Left of drain	Right of drain	Left of drain	Right of drain
Near the drain	136	124	132	128	120	110
Intermediate section	120	113	117	106	124	83
Away from the drain	124	98	113	85	106	72

Note: Values are in kPa.

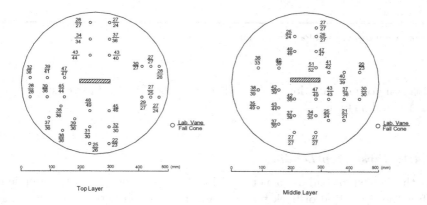

Fig. 5.13. Measured undrained shear strength in kPa after compression.

higher near the drain. The middle layer has higher strength than the top layer which is consistent with the higher moisture content found in the top layer. Measured strengths varied between 20 and 47 kPa based on the laboratory vane tests, and between 21 and 49 kPa based on fall cone tests. These values are more or less similar to the undrained strengths computed from the effective stress using the correlation proposed by Ladd *et al.* (1977). The comparison of computed and measured undrained shear strength at various locations is shown in Table 5.6. The above findings indicate that the sample was not consolidated homogeneously. More improvements were found near the drain.

5.4. Incremental Step Loading Test

The incremental step loading tests were carried out on the high moisture content ultra-soft slurry-like soil to study the deformation characteristics

Table 5.6. Comparison of c_u measured in the laboratory and calculated applying study of Ladd et al. (1977) for two-step high pressure loading test.

	Top layer				Middle layer				
	Left of drain		Right of drain		Left of drain		Right of drain		
	A	B	A	B	A	B	A	B	Remarks
Near the drain	43	38	45	35	47	34	47	24	$c_u/p' = 0.28$
Intermediate section	34	34	31	32	27	37	—	—	
Away from the drain	28	35	26	27	27	33	27	22	

Note: A = measured in kPa; B = computed in kPa.

of this material under low magnitude of loading. The load increments were 12, 19, and 44 kPa.

5.4.1. First Step Loading

In the first step loading, a net additional pressure of 12 kPa was applied after the deduction of back pressure and the friction between the loading plate and the cylinder wall. The pressure cell registered 22 kPa at the top and 30 kPa at the bottom. The sample thickness was about 700 mm with a total bulk density of 12.85 kN/m^3. The measured pore pressures were slightly different at the top and bottom due to difference in the hydrostatic pressures. However, the measured pore pressures tallied with the total applied pressure plus hydrostatic pressure. The first load step was carried out for 30 days. No pore pressure dissipation was registered at all transducers throughout the tests, as shown in Fig. 5.14. The total pressures measured from both the top and bottom cells were constant throughout the test, as shown in Fig. 5.15.

During the first step loading 54 mm of settlement took place. This is equivalent to 8% of the strain occurring without pore pressure dissipation. The void ratio at the end of the first step is about 4.21. The period of no pore pressure dissipation in this case exceeded 30 days which was much longer than that with the two-step high pressure loading which took 6–10 days. This could be due to the low pressure gradient which retarded the flow rate.

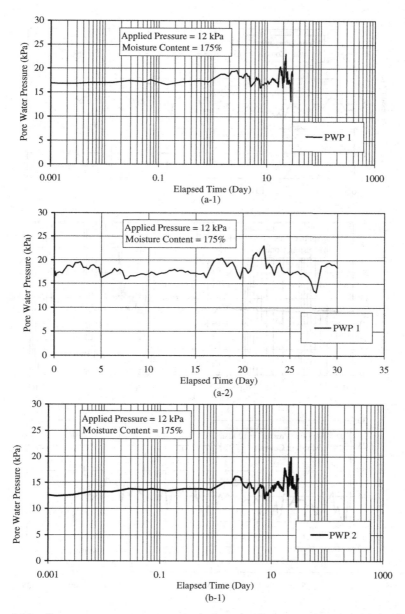

Fig. 5.14. Pore pressure measurement during the first step incremental loading: (a) PWP 1; (b) PWP 2; (c) mini-PWP 1; (d) mini-PWP 2; (e) mini-PWP 3; (f) mini-PWP 4. (1 and 2 refer to semi-log scale and arithmetic scale, respectively.)

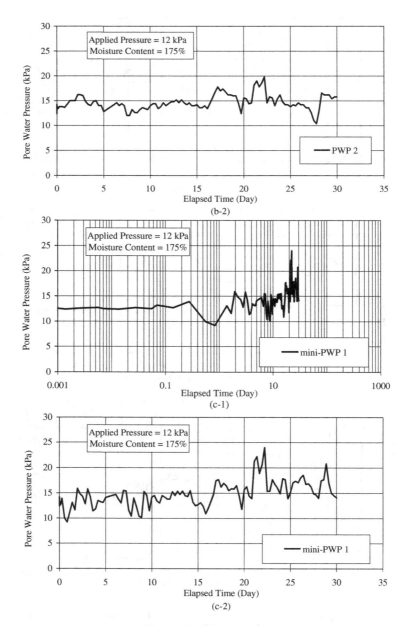

Fig. 5.14. (*Continued*).

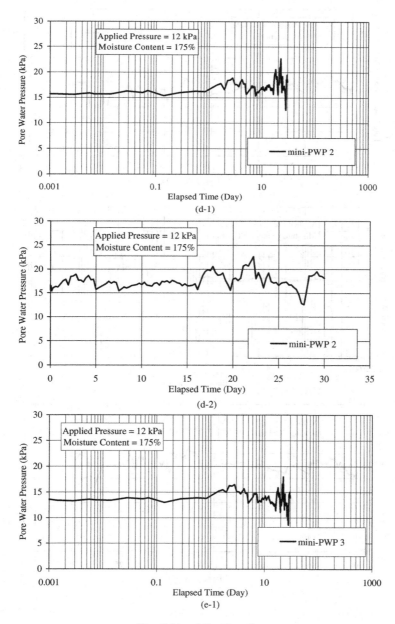

Fig. 5.14. (*Continued*).

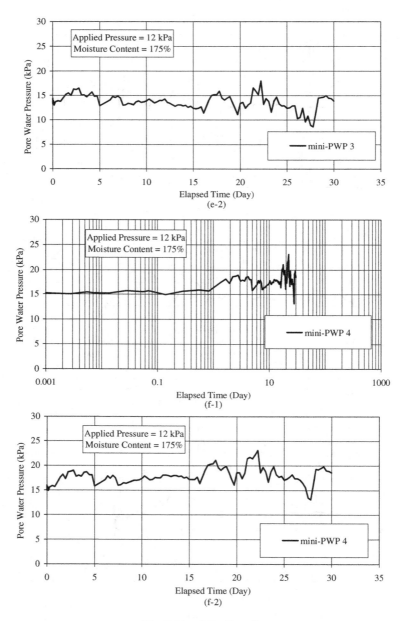

Fig. 5.14. (*Continued*).

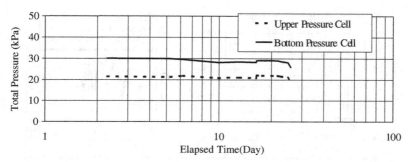

Fig. 5.15. Total pressure measurements at the top and bottom pressure cells during the first step incremental loading.

5.4.2. Second Stage Loading

In the second step loading, a net pressure of 19 kPa was applied to the soil. Since no pore pressure dissipation had occurred in the first stage, an increase in the pore pressure of 19 kPa was measured at the start of the second stage (Table 5.7). It should be noted that the measured pressure is total pore pressure and not excess pore pressure.

The second stage loading was carried out up to 110 days. Again, no pore pressure dissipation was observed in all transducers although 145 mm of settlements (21% strain) was measured, as shown in Figs. 5.16(a)–5.16(f). Occasionally, the pore pressure dropped within one day, however, it picked up again the next day. This could be due to fluctuation of loading as shown

Table 5.7. Summary of measurements in step incremental loading.

	12 kPa loading		19 kPa loading		44 kPa loading		Calculated effective stress gain (kPa)
	Initial (kPa)	Final (kPa)	Initial (kPa)	Final (kPa)	Initial (kPa)	Final (kPa)	
Mini-PWP 1	14	14	19	17	—	—	—
Mini-PWP 2	17	18	23	23.6	49.8	28	24
Mini-PWP 3	14	14	19	17	44.6	27	20
Mini-PWP 4	16	19	23	22	50.2	27	25
PWP 1	16	18	23.4	23.8	51.6	37.6	16
PWP 2	14	16	20	21	48.8	35	16
Pressure cell (Upper)	22	21	27	26	52	60	—
Pressure cell (Bottom)	30	26	34	32	57	70	—
Air pressure	22	22	32	32	57	57	—

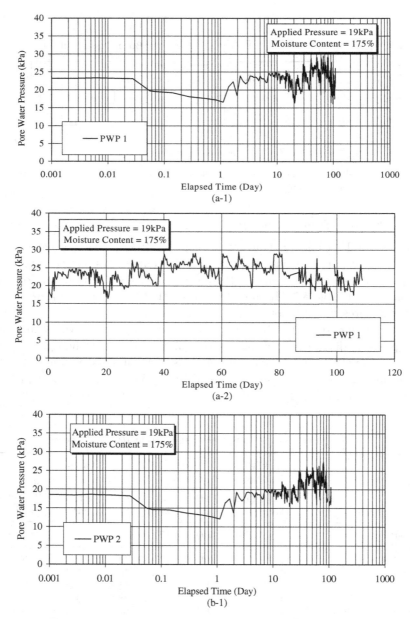

Fig. 5.16. Pore pressure measurement during the second step incremental loading: (a) PWP 1; (b) PWP 2; (c) mini-PWP 1; (d) mini-PWP 2; (e) mini-PWP 3; (f) mini-PWP 4. (1 and 2 refer to semi-log scale and arithmetic scale, respectively.)

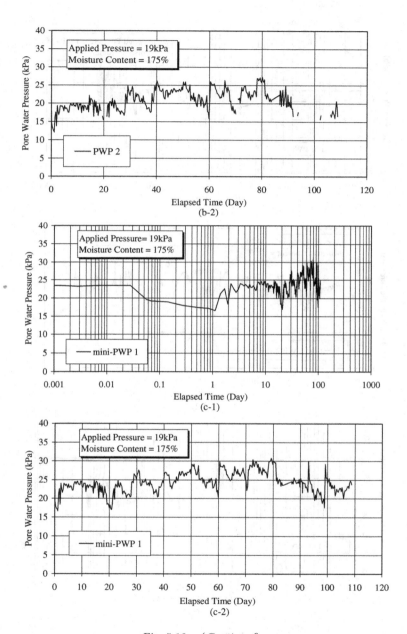

Fig. 5.16. (*Continued*).

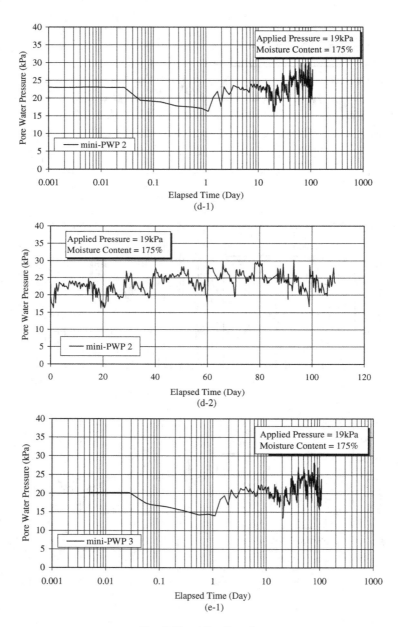

Fig. 5.16. (*Continued*).

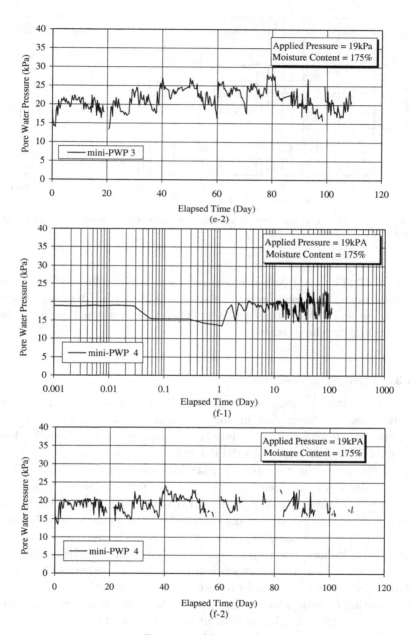

Fig. 5.16. (Continued).

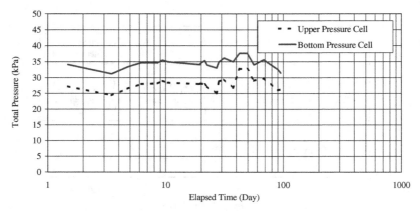

Fig. 5.17. Total pressure measurements at the top and bottom pressure cells during the second step incremental loading.

in Fig. 5.17. The total pressure measured was slightly fluctuated and no friction built-up was measured in this stage (Fig. 5.17). The average void ratio at the end of this stage was 3.04. Therefore, it seems that ultra-soft soil is still in the slurry state.

5.4.3. Third Step Loading

In the third step loading, net applied pressure was 44 kPa. Since there was no pressure dissipation in the first and second stages, the excess pore pressure measured was essentially the same as the total applied load (Table 5.7). Except for the mini-PWP 4, other transducers did not show any signs of dissipation up to 5–10 days after loading (Figs. 5.18(a)–5.18(e)).

At mini-PWP 4 location, the excess pore pressure dissipated immediately while mini-PWP 2, 3, and 5 took 10 days and PWP 1 and 2 took 8 days. At the time of initiation of pore pressure dissipation, the total settlement were 260, 283, and 290 mm, and the average void ratios were 2.54, 2.36, and 2.3, respectively. Therefore, the transition point for the slurry becoming soil corresponds to a void ratio between 2.3 and 2.6 which is very close to the void ratio at the liquid limit of 2.28. Tests were carried out up to 30 days in the third stage. The dissipation of pore pressure was only 10–20 kPa. No friction built-up was registered throughout the tests (Fig. 5.19).

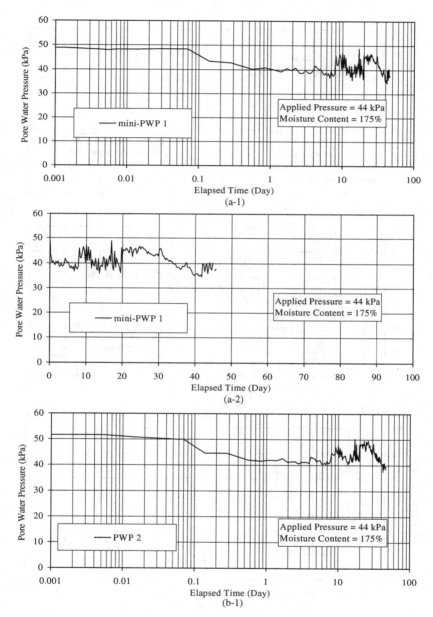

Fig. 5.18. Pore pressure measurement during the third step loading: (a) mini-PWP 1; (b) PWP 2; (c) mini-PWP 2; (d) mini-PWP 3; (e) mini-PWP 4. (1 and 2 refer to semi-log scale and arithmetic scale, respectively.)

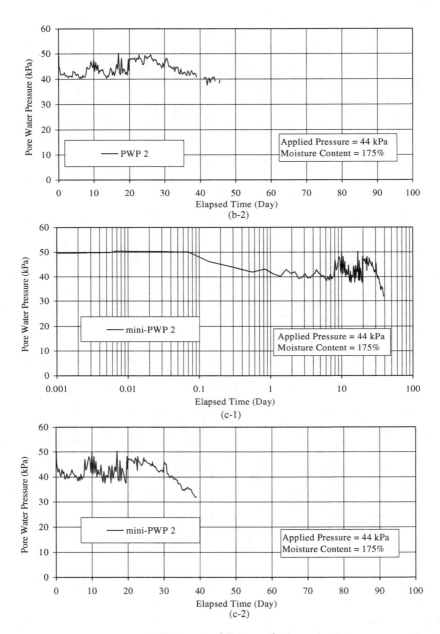

Fig. 5.18. (Continued).

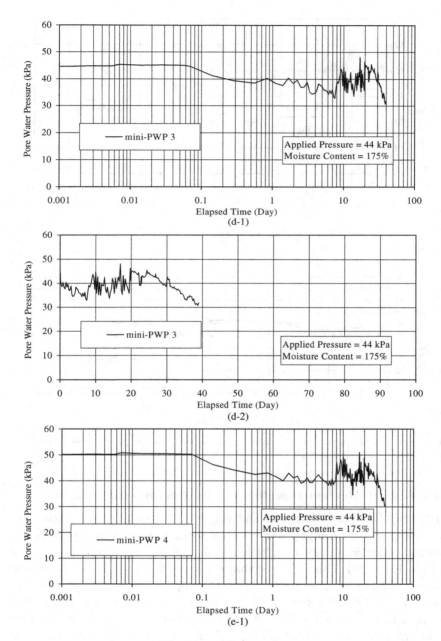

Fig. 5.18. (*Continued*).

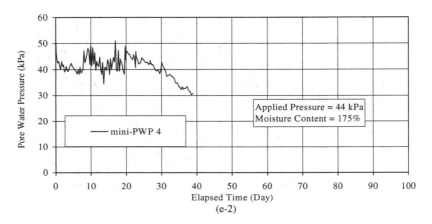

Fig. 5.18. (*Continued*).

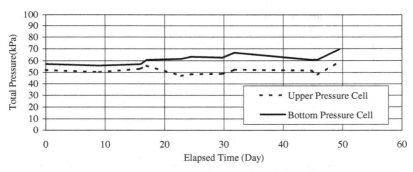

Fig. 5.19. Total pressure measurements at the top and bottom pressure cells during the third step incremental loading.

5.4.4. Post-Mortem Investigation on Compressed Sample

The clay block was extruded and cut open for close examination after completion of the compression test. The bottom group of mini-piezometers moved down vertically by 119 mm and laterally by 75 mm toward the drain (Fig. 5.20). This indicated that the bottom 240 mm layer compressed about 119 mm or a strain of 49%. The top group of mini-piezometers settled 260 mm. Therefore, a 320 mm thick layer between the two groups of piezometers settled 141 mm which was about 44% strain. The topmost 140 mm thick layer settled 71 mm which was about 51% strain. Again the highest strain occurred at the top layer, followed by the lowest

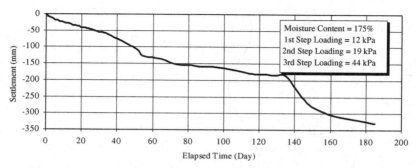

Fig. 5.20. Vertical settlement measurement during the first, second, and third step high pressure loading.

layer, and finally the middle layer. The final void ratios worked out from sublayer settlement varied between 1.77 and 2.16. Details are shown in Table 5.4.

Laboratory tests were carried out on samples obtained from the clay block. Again, the migration of fines toward the drain was evident. The moisture contents were low near the drain and increased with distance from the drain. The moisture content varied between 70.4 and 93.17, and the corresponding void ratios were 1.89–2.5, as shown in Fig. 5.21. This variation is slightly different from the void ratios computed from

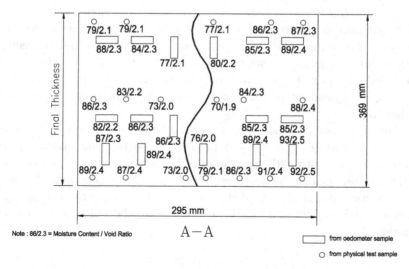

Fig. 5.21. Measured moisture content and void ratio after compression test.

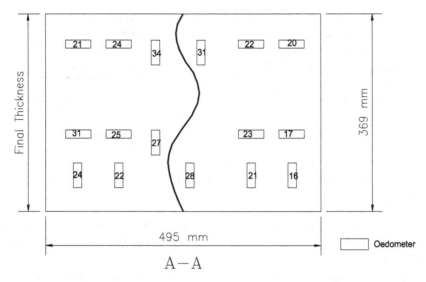

Fig. 5.22. Measured preconsolidation pressure in kPa after compression test.

the settlements of piezometer tips. The preconsolidation pressures varied between 16 and 34 kPa with higher values near the drain, as shown in Fig. 5.22. The measured preconsolidation pressures are much lower than the final applied pressure. This is because the tests were not carried out to the end of primary consolidation and in the first two steps, there was no significant gain in effective stress. The measured preconsolidation pressures were even lower than the second stage loading of 19 kPa. The measured preconsolidation pressures were compared with the effective stress gain calculated based on the pore pressure dissipation in the third stage. This was in good agreement because there was no friction built-up in the test (Table 5.7).

The undrained shear strength measured by laboratory vane test varied between 7.6 and 18, and 9.2 and 14 kPa for the top and middle layers, respectively. The strength decreases with increasing distance from the vertical drain as can be seen in Fig. 5.23. The relationship between moisture content and undrained shear strength measured after compression is shown in Fig. 5.24 and can be expressed by the following regression equation:

$$S_u = 78.128 - 0.828\,W\,\text{kPa} \qquad (5.1)$$

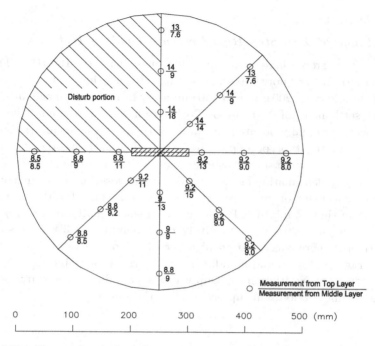

Fig. 5.23. Measured undrained shear strength in kPa by lab vane after compression.

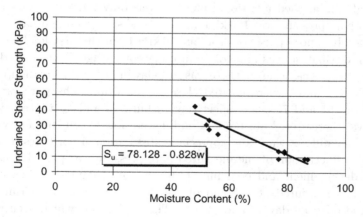

$S_u = 78.128 - 0.828w$

Fig. 5.24. Correlation between moisture content and undrained shear strength.

5.5. Summary

First Stage of Two-Step High Pressure Loading Test

In the first step there was an additional load of 100 kPa. There was essentially no pore pressure dissipation in the first 10 days at piezometers located 200 mm and 250 mm away from the drainage although a total settlement of 150 mm or 20% strain was measured. The two transducers originally located 100 mm away from the drain registered the commencement of pore pressure dissipation at half-a-day at mini-PWP 1 and 6 days at mini-PWP 2. Lateral soil movement during the initial stage of deformation brought mini-PWP 1 closer to the drain causing the commencement of pore pressure dissipation at mini-PWP 1 to occur sooner than that at mini-PWP 2. The settlement at half-a-day and 6 days were 35 mm and 120 mm, respectively corresponding to 3%–10% strain, respectively. There was a period of no or little pore pressure dissipation. The average void ratio and liquidity index at transition point are 2.68 and 1.6, respectively. No friction built-up was observed during the slurry stage, but significant friction built-up occurred at the soil stage.

Second Stage of Two-Step High Pressure Loading Test

Although the applied additional pressure was only 90 kPa, the measured induced pore pressure was 120 kPa in the second step loading because there was a residual pore pressure of about 30 kPa from the first loading. Pore pressure dissipation was immediate in the second stage. This showed that when a slurry became a soil, there was no delay in pore pressure dissipation. Mini-PWP 1 and 2 were located close to the drain. The pore pressure dropped to 57 and 75 kPa, respectively within 24 h and fluctuated around that level thereafter. Mini-PWP 3 and 4 were located further away from drain. They did not show any appreciable dissipation of pore pressure with time. The fluctuation of the measured pore pressure was probably caused by gas development and cracking of the soil. A maximum settlement of 13 mm was measured at the end of day 17 and swelling of 0.5 mm was registered between day 17 and day 23. The settlement, which occurred in the second stage was much smaller than that occurred in the first stage. The gain in effective stress was higher near the drain and lower at locations away from the drain. The preconsolidation pressure after the second stage varied between 85 and 136 kPa. With the total applied load of 190 kPa the measured undrained shear strength varied between 20 and 49 kPa at the end of the second stage loading.

Incremental Step Loading Tests

Large compression occurred in the ultra-soft slurry under a small applied pressure. The no-pore pressure dissipation period was much longer due to the low hydraulic gradient. The gain in effective stress in the third step under 44 kPa applied load was only 16–34 kPa at the end of 30 days. An undrained shear strength of 8–14 kPa was achieved after the third step loading.

Chapter 6

COMPRESSION TEST ON SLURRY WITH SMALL-SCALE CONSOLIDOMETER

The large-scale test generally takes too long to complete and also costly even for a one-step loading. It was, therefore, not suitable to investigate the variation of compression behavior in the slurry stage due to variation of initial moisture contents. To understand the compressibility characteristics of the ultra-soft soil in the slurry-like state, compression tests were carried out with a small-scale consolidometer, which is equipped with pore pressure transducers at three different locations. The deformation behavior was monitored with both settlement and pore pressure measurements. The possible transition points were determined from the settlement rate, change in void ratio, change in hydraulic conductivity, and pore pressure dissipation.

6.1. Description of the Apparatus

The small-scale consolidometer consisted of a copper cylinder with an inside diameter of 89.5 mm and an outside diameter of 99.5 mm. The cylinder sat on the triaxial pedestal which was fitted with pore pressure and back-pressure lines. The top plate or piston was indented at the center to accommodate the loading rod. The piston had two drainage holes and was fitted with a porous stone filter. Therefore, drainage was permitted only at the top. The friction between the piston and the cell was negligible since the piston could fall freely when there was no sample in the cell. The total height of the cylinder was 224 mm and the sample was prepared to an initial height of up to 150 mm. Two additional pore pressure transducers were fixed at the side of the cylinder at 37 and 74 mm above the bottom. Direct loading can be applied to the piston through a loading hanger. Features of the small-scale consolidometer are shown in Fig. 6.1.

6.2. Method of Testing

To understand the compression characteristics of the ultra-soft soil upon loading, one-dimensional compression tests were carried out on 150 mm height samples with four different moisture contents of 130%, 150%, 170%, and 190%. Each sample was subjected to a one-step applied load. Three different applied loads were used at 127, 175, and 223 kPa. The settlement and pore pressures were recorded throughout each test.

6.3. Discussion on Test Results

Both settlement and pore pressures were measured under the one-dimensional K_0 consolidation tests. Immediately after loading, the settlement commenced. The measured pore pressures were essentially the same or slightly higher than the applied pressures. No pore pressure dissipation was registered in all three piezometers for a long time. Even though one of which is only about 76 mm away from the drainage boundary.

6.3.1. Settlement Response

6.3.1.1. Effect of Initial Moisture Content

To analyze the deformation behavior of ultra-soft slurry-like soil upon the application of an additional load, the settlement behavior of soil with different initial moisture contents under the same load was studied. The settlement curves of samples with different initial moisture contents (130%, 150%, 170% and 190%), under the applied loads of 127, 175 and 223 kPa are shown in Figs. 6.2–6.4. The results show that the total settlement increases with increasing initial moisture content.

The rate of settlement was faster in the early stage of test. During the early stage, the settlement rate increases with increasing initial moisture content. After a certain time, all the curves converge. During the early stage of settlement, the rate of settlement was reducing with time. The rate of reduction of the settlement rate increases with the increase in initial moisture content of the ultra-soft soil. After reaching the convergence point,

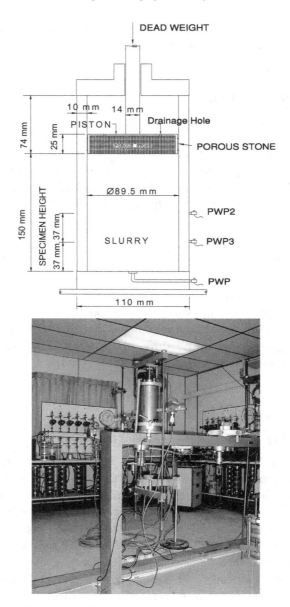

Fig. 6.1. Small-scale consolidometer equipped with pore pressure transducers.

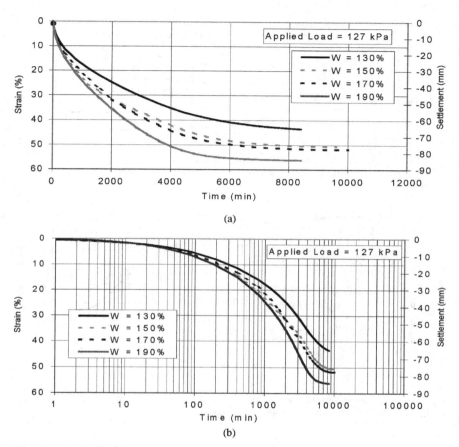

Fig. 6.2. Settlement (strain) versus time for various moisture contents under the same loading of 127 kPa. (a) arithmetic scale, (b) log scale.

the rate of settlement decreased rapidly. In this case convergence point is a point where all the settlement rate graphs converged into one point. Curves of settlement rate versus time for various moisture contents under various magnitudes of one-step loading are shown in Figs. 6.5(a)–6.5(c).

The convergence point was found to occur at about 4200, 3400, and 2940 min in the 127, 175, and 223 kPa loadings, respectively (Table 6.1). The void ratio at the convergence point appears to be independent of initial moisture content. After this point, the rates of settlement were slightly different for various initial moisture contents. Therefore, this is an easy way to determine the transition point with only one test measuring

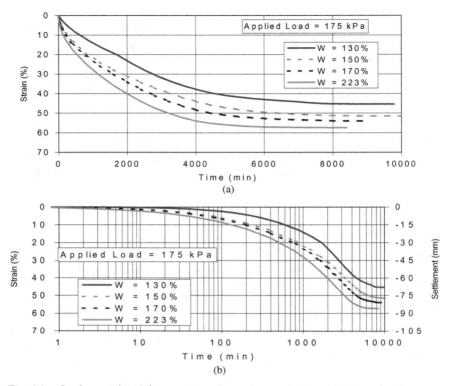

Fig. 6.3. Settlement (strain) versus time for various moisture contents under the same loading of 175 kPa. (a) arithmetic scale, (b) log scale.

the settlement rate. The point of commencement of rapid decrease in settlement rate will indicate the approximate point of changing from slurry-like behavior to soil-like behavior. It is also noted that at the point of convergence, the strain rate was 0.01%/min or about 0.01 mm/min for sample with an initial thickness of 150 mm. Curves of strain rate versus time for various moisture contents under various loadings are shown in Figs. 6.6(a)–6.6(c).

(i) Void Ratio Change

The highest void ratio changes occurred at an early stage. The higher the initial void ratio, the greater is the magnitude and rate in void ratio changes. The magnitude of void ratio changes in the first 10 min is insignificant. The void ratios of all samples with different moisture contents converged into

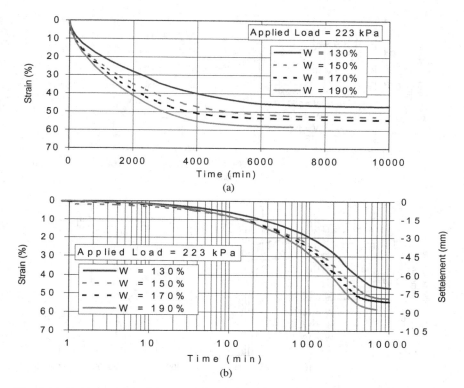

Fig. 6.4. Settlement (strain) versus time for various moisture contents under the same loading of 223 kPa. (a) Natural scale, (b) log scale.

one at void ratios between 1.9 and 2. The curves of void ratio versus time for various moisture contents under various magnitudes of loading are shown in Figs. 6.7(a)–6.7(c). A comparison of the time and void ratios at the converging point for various magnitudes of loading are shown in Table 6.2.

From the above observations the following conclusions can be made on the deformation behavior of the ultra-soft soil with different moisture contents under the same magnitude of loading:

- Samples with various initial moisture contents eventually settled to the same void ratio under the same load.
- The settlement rate or rate of void ratio change is higher in the early stage when the material is still in a slurry state.

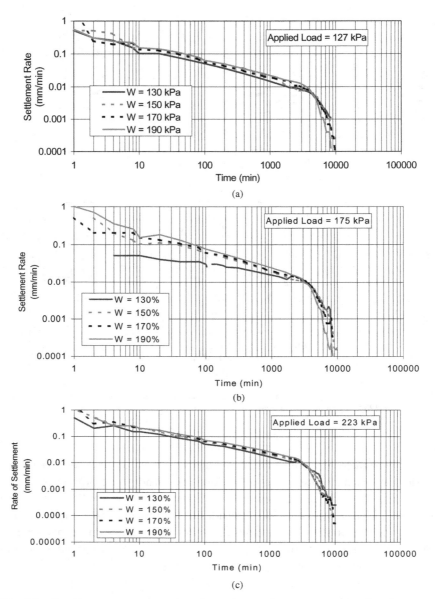

Fig. 6.5. Settlement rate versus time for various moisture contents under the same loading. (a) 127 kPa, (b) 175 kPa, (c) 223 kPa.

Table 6.1. Convergent point based on settlement rate.

	Applied pressure (kPa)		
	127	175	223
Elapsed time at convergence point	4217	3400	2942
Void ratio convergence point	1.92	2	1.9
Liquidity indices at convergence point	0.71	0.76	0.70

Note: Values are same for all moisture content soils.

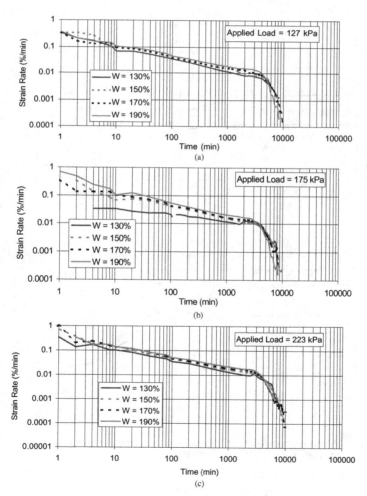

Fig. 6.6. Strain rate versus time for various moisture contents under the same loading. (a) 127 kPa, (b) 175 kPa, (c) 223 kPa.

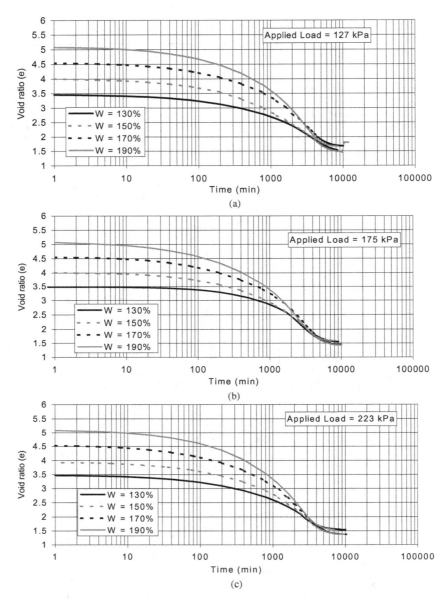

Fig. 6.7. Void ratio versus time for various moisture contents under the same magnitude of load. (a) 127 kPa, (b) 175 kPa, (c) 223 kPa.

Table 6.2. Time and void ratio at convergence determined from merging of void ratio curves.

Applied pressure (kPa)					
127		175		223	
Elapsed time	Void ratio	Elapsed time	Void ratio	Elapsed time	Void ratio
4375	1.92	3162	2.0	2966	1.9

- At one point of time, the settlement rates as well as the void ratios converge. After that point, a rapid decrease in the change in void ratio was observed.
- The convergence point is believed to be the transition point between viscous state and elastoplastic state. This would be compared with transition points determined by other methods. This point occurred at a void ratio of about 1.9–2 and the strain rate at that point is about 0.1%/min or 0.01 mm/min. It took a longer time to reach the transition point with a load of lower magnitude. However, it is interesting to note that samples of various moisture contents under the same magnitude of load reach the interception point at almost the same time.

(ii) Hydraulic Conductivity Change

Been and Sills (1981) proposed that the hydraulic conductivity can be calculated from the solid velocity using the following equation:

$$k = -\nu_s/i \tag{6.1}$$

where ν_s is the solid velocity and i is the hydraulic gradient.

The solid velocity is simply the rate of change in sample height and the hydraulic gradient can be determined from the pore pressure measurements. The average hydraulic conductivity of a sample can, therefore, be calculated using the settlement and pore pressure data at any time.

It can be seen in Figs. 6.8(a)–6.8(c) that the hydraulic conductivity variations are higher in the early stage. The higher the initial moisture content, the greater is the hydraulic conductivity. The hydraulic conductivity reduces with time and the rate of reduction of hydraulic conductivity is greater for samples of higher moisture content since these samples settle at a faster rate. At a certain point in time, the hydraulic conductivities of the different samples converge into one and the rate of hydraulic conductivity reduction was changed. This point is likely to be the transition point from ultra-soft soil to structured soil. At this point,

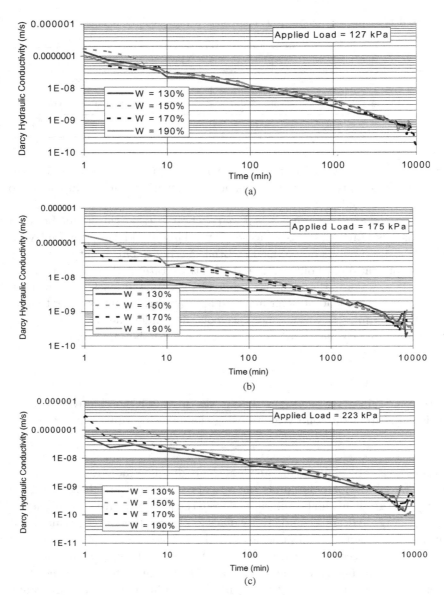

Fig. 6.8. Darcy hydraulic conductivity versus time for various moisture contents under the same magnitude of loading. (a) 127 kPa, (b) 175 kPa, (c) 223 kPa.

Table 6.3. Time and void ratio at convergence determined from hydraulic conductivity of ultra-soft soils.

Applied pressure (kPa)					
127		175		223	
Elapsed time	Void ratio	Elapsed time	Void ratio	Elapsed time	Void ratio
4548	1.83	2976	2.1	2801	2.0

the hydraulic conductivity is around 10^{-9} m/s. The time of convergence is similar to those obtained from void ratio changes and from settlement analysis (Table 6.3).

(iii) Hydraulic Conductivity Change Index, C_k

The hydraulic conductivity change index is a measure of the change in void ratio with respect to the change in hydraulic conductivity, as defined in Eq. (6.2).

$$C_k = -\Delta e / \log(k_2/k_1) \qquad (6.2)$$

where C_k is the hydraulic conductivity change index, Δe is the void ratio change, and k_n is the hydraulic conductivity at certain void ratio.

The curves of hydraulic conductivity versus void ratio for various initial moisture contents (Fig. 6.9) under the same loading showed that curves are non-linear. Each curve can be divided into three segments. In the first segment, the change in hydraulic conductivity is very high with minimum changes in void ratio. At this stage the migration of fines would have occurred with minimum volume of water drained out from the cell. Fine materials might have been migrated toward the drainage boundary and started forming a filter mud cake on the porous stone. Therefore, the hydraulic conductivity started decreasing due to rearrangement of the soil grains. The second segment shows significant variations of hydraulic conductivity change indices with void ratio reduction. In the third segment the same hydraulic conductivity change index was noted for all moisture contents. The beginning of the third segment is likely to be the point where the ultra-soft slurry-like soil began behaving like the structured soil. The hydraulic conductivity change indices at the third stage of all the soil with different initial moisture contents are found to be similar even under different loads. However, the corresponding void ratios are found to be in the range of 2.5–3, which are higher than those determined from settlement analysis and the rate of hydraulic conductivity changes.

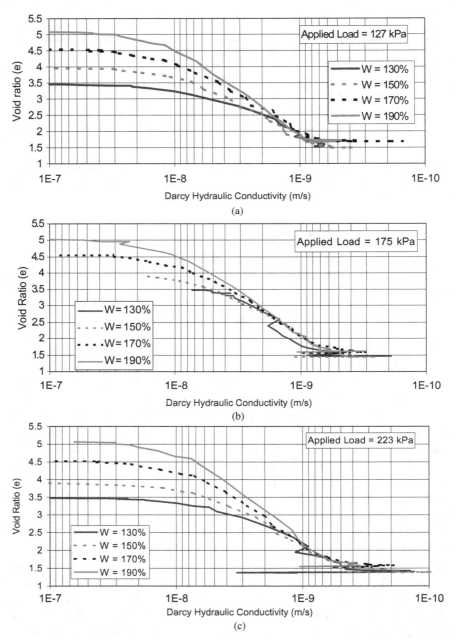

Fig. 6.9. Void ratio versus Darcy hydraulic conductivity for various moisture contents under the same loading. (a) 127 kPa, (b) 175 kPa, (c) 223 kPa.

Different hydraulic conductivity change indices C_k were found at the different stages. At the early stage, C_k ranges from 0.04 to 0.2. Tavenas et al. (1983a, b) and Leroueil (1999) have reported that C_k is approximately equal to $0.5e_0$ for normal soil. These hydraulic conductivity change indices are far lower than the value they reported. At the second stage, C_k increases with increasing initial water content. Change in hydraulic conductivity with void ratio for samples of various initial moisture contents are shown in Figs. 6.9(a)–6.9(c). Samples with the same initial moisture content under various loadings yield similar hydraulic conductivity change index values as shown in Figs. 6.10(a)–6.10(c).

The hydraulic conductivity change index in the second stage ranges between 0.97 and 2.5. Figure 6.11 shows a correlation between the hydraulic

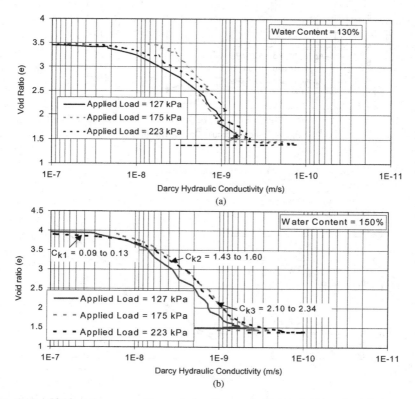

Fig. 6.10. Void ratio versus Darcy hydraulic conductivity for samples of the same moisture content soil but under various loadings. (a) $W = 130\%$, (b) $W = 150\%$, (c) $W = 170\%$, (d) $W = 190\%$.

Compressibility of Ultra-Soft Soil

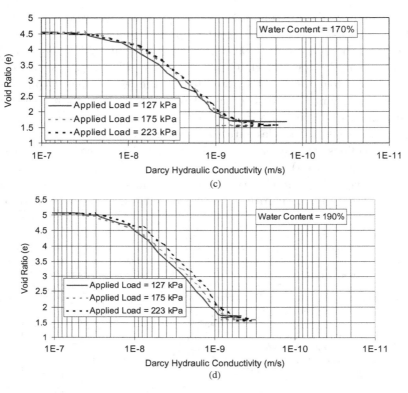

Fig. 6.10. (*Continued*).

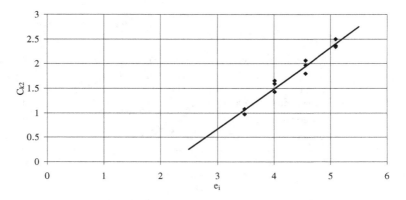

Fig. 6.11. Hydraulic conductivity change index versus initial void ratio.

Table 6.4. Hydraulic conductivity change indices of various moisture content soils at various stages.

Moisture content (%)	Applied pressure (kPa)								
	127			175			223		
	1st Stage	2nd Stage	3rd Stage	1st Stage	2nd Stage	3rd Stage	1st Stage	2nd Stage	3rd Stage
130	0.114	0.97	2.1	—	—	2.34	0.13	1.08	2.2
150	0.09	1.6	2.1	—	1.66	2.34	0.13	1.43	2.2
170	0.06	1.97	2.1	—	2.06	2.34	0.15	1.8	2.2
190	6.04	2.5	2.1	—	2.34	2.34	0.2	2.37	2.2

conductivity change index at the second stage with initial void ratio.

$$C_k = 0.83e_i - 1.82 \qquad (6.3)$$

However, this may be valid only for high moisture content soils. The hydraulic conductivity change indices fall into the narrow range between 2.1 and 2.34 in the third stage. The ultra-soft soil becomes a structured soil at more or less the same void ratio. Table 6.4 shows hydraulic conductivity change indices at different stages for various moisture contents under various loadings.

It should be noted that for the deformation at the early stage under the same magnitude of loading, the same hydraulic conductivity values were measured for different void ratios. The hydraulic conductivity decreases rapidly with a minimum reduction in void ratio. This phenomena was explained by Tavenas et al. (1983a) for the underestimation of the hydraulic conductivity change index. This is because the average void ratio is larger than the hydraulic conductivity of the soil near the drainage boundary. At this early stage, it was likely that the flow rate was reducing due to migration of soil particles toward the porous stone. At the second stage, it was also surprising to observe that for the same void ratio, different hydraulic conductivity values were observed. A similar condition was also reported by Katagiri and Imai (1994) and Katagiri (1995) in the study on void ratio hydraulic conductivity relationship for high moisture content soils. It could be due to the same reason that the hydraulic conductivity was mainly controlled by soil close to the porous stone, whereas the void ratio was taken as an average of the entire sample. Therefore, the relationship of the hydraulic conductivity changes with void ratio from these test including the rate of settlement which was controlled by hydraulic conductivity, was considered to be only qualitative and not sufficiently accurate for back analyses. To eliminate this problem, the tests were repeated on thinner samples.

6.3.1.2. *Effect of Magnitude of Applied Pressure*

The compression behavior of the ultra-soft soils with the same initial moisture content but under different applied pressures was studied. Tests were conducted for samples with moisture content of 130%, 150%, 170%, and 190% and applied pressures of 123, 175, and 223 kPa.

It can be seen in Figs. 6.12–6.15 that the difference in magnitude of settlement for samples with the initial same moisture content subjected to the different loads was not significant although a higher loading resulted in

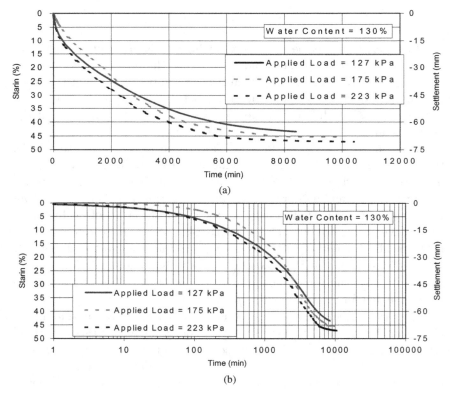

Fig. 6.12. Settlement (strain) versus time for samples of the same moisture content but under various loadings. (a) arithmetic scale, (b) log scale.

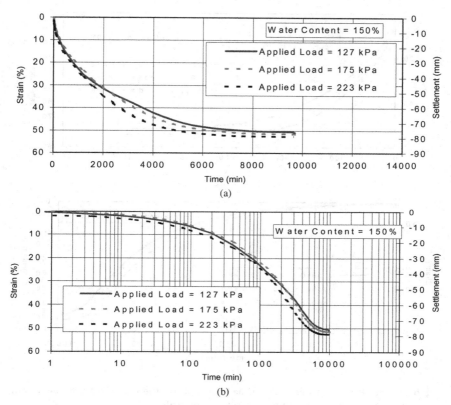

Fig. 6.13. Settlement (strain) versus time for samples of the same moisture content but under various loadings. (a) arithmetic scale, (b) log scale.

slightly higher settlements. The higher the magnitude of loading the lower is the final void ratio (Fig. 6.16). The initial and final void ratios of samples with various initial moisture contents under various loadings are shown in Table 6.7. Therefore, it can be concluded that the variation of settlement or void ratio changes mainly occurred in the slurry stage and they were mainly caused by the difference in initial moisture content rather than the magnitude of loading.

Figure 6.16 explains the difference in settlement magnitude due to variations in the initial void ratio and magnitude of loading. It can be seen in Figs. 6.17(a)–6.17(d) that the rate of settlement were more or less the same in both the slurry state and soil state under various magnitudes of load. However, at a certain time there was a kink in the change of settlement

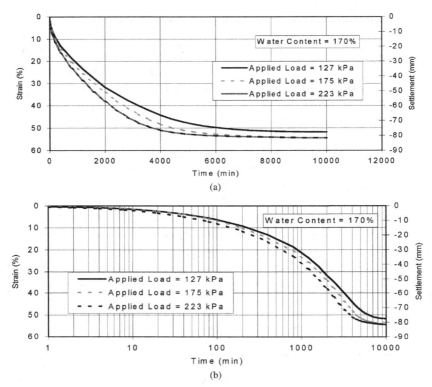

Fig. 6.14. Settlement (strain) versus time for samples of the same moisture content but under various loadings. (a) arithmetic scale, (b) log scale.

rate. After the kink, the settlement rate decreased rapidly. However, the transition points for the samples of the same initial moisture content but under different loads are slightly different. The time required to reach the transition points are shown in Table 6.1. It can be seen that a longer time was required to reach the transition point for lower magnitudes of loading. The curves of strain rate versus time are shown also in Figs. 6.18(a)–6.18(d).

It was also found that for the same initial moisture content soil, the rate of change in void ratio with time is only slightly different under various loadings. A higher loading gave slightly faster void ratio changes (Figs. 6.19(a)–6.19(d)).

Comparison of hydraulic conductivity was also made on the same initial moisture content soil under various loadings. It was found that for the same time duration, the hydraulic conductivity under low magnitude of

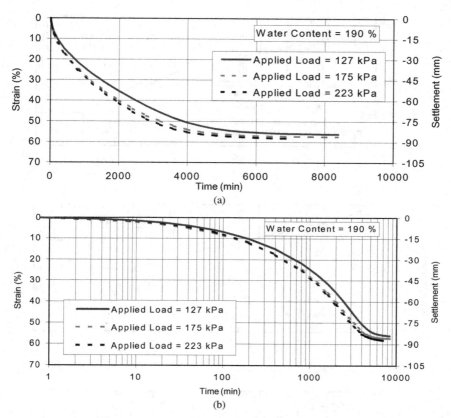

Fig. 6.15. Settlement (strain) versus time for samples of the same moisture content but under various loadings. (a) arithmetic scale, (b) log scale.

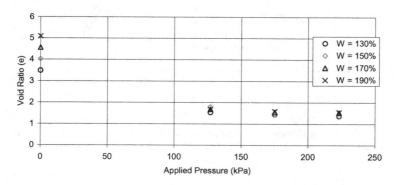

Fig. 6.16. Void ratio versus applied pressure from various compression tests.

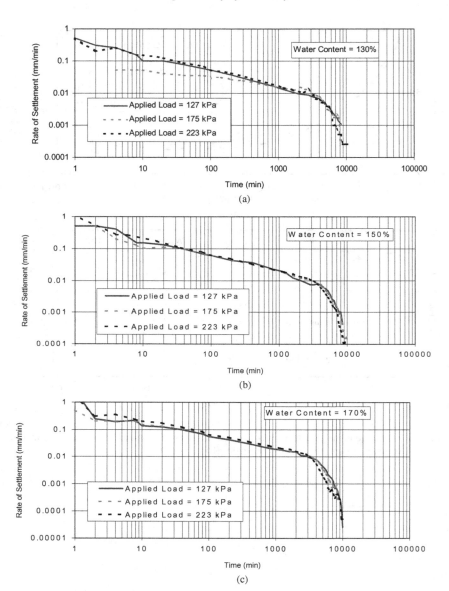

Fig. 6.17. Rate of settlement versus time for samples of the same moisture content but under various loadings. (a) $W = 130\%$, (b) $W = 150\%$, (c) $W = 170\%$, (d) $W = 190\%$.

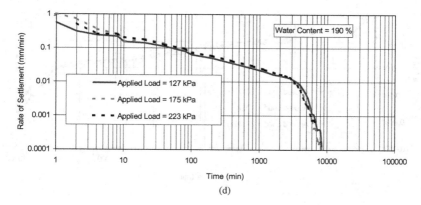

(d)

Fig. 6.17. (*Continued*).

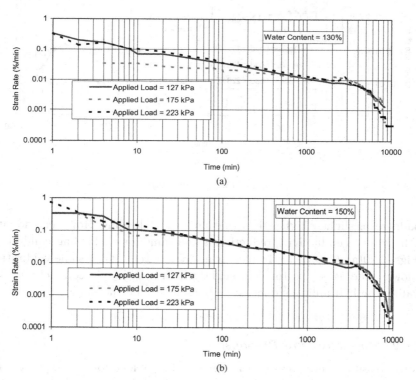

(a)

(b)

Fig. 6.18. Strain rate versus time for samples of the same moisture content but under various loadings. (a) $W = 130\%$, (b) $W = 150\%$, (c) $W = 170\%$, (d) $W = 190\%$.

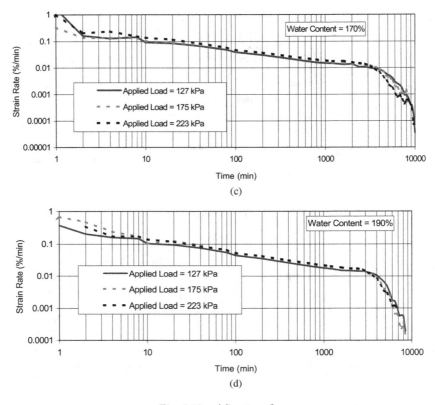

Fig. 6.18. (*Continued*).

loading was slightly higher than that under high loading. However, the rate of hydraulic conductivity reduction is the same for all loading conditions (Figs. 6.20(a)–6.20(d)). It is interesting to note that at the same void ratio, a lower loading seems to yield slightly higher hydraulic conductivity. In reality, the hydraulic conductivity values should be the same.

6.3.2. *Pore Pressure Response*

Pore pressures were measured at three different locations as explained in the earlier chapter. The data was analyzed based on measurements at the center of the bottom of the sample.

6.3.2.1. Effect of Initial Moisture Content

Pore pressure responses in samples with various initial moisture contents subjected to the same magnitude of loading were studied. It can be seen in Figs. 6.21(a)–6.21(c) that the pore pressure responded in all cases within a minute after the application of loads. The pore pressures were about the same or slightly higher than the applied pressure for more than 1000 min. The time to the commencement of pore pressure dissipation were different for the various initial moisture contents under the same load. It could be due to the higher settlement and strain rate in the higher moisture content soil which resulted in a shorter drainage path and hence a faster rate of pore pressure dissipation.

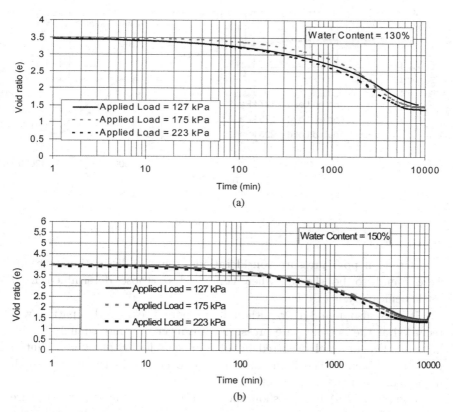

Fig. 6.19. Void ratio versus time for sample of the same moisture content but under various loadings. (a) $W = 130\%$, (b) $W = 150\%$, (c) $W = 170\%$, (d) $W = 190\%$.

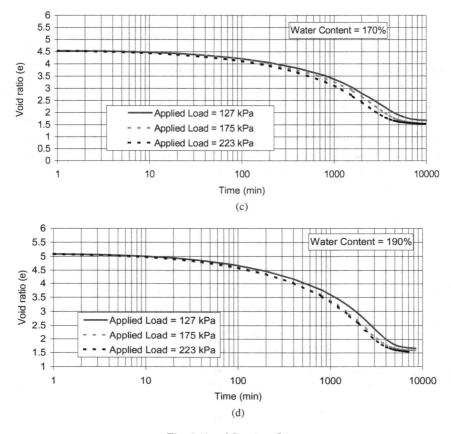

Fig. 6.19. (*Continued*).

Based on pore pressure considerations, the "no dissipation" periods were found to be shorter when compared to those determined from settlement analysis (Table 6.5).

It was also found that although no pore pressure dissipation was recorded at the bottom pore pressure transducer for more than 1000 min, large settlements had occurred in all four samples with different initial moisture contents. The highest settlement occurred in the sample with highest initial moisture content.

Figures 6.22(a)–6.22(c) show the variation of measured pore pressure with void ratio for samples of various initial moisture contents under the same magnitude of loading. It is obvious that void ratios were changing without the significant dissipation of pore pressure and commencement of pore pressure dissipation started earlier in samples of high moisture content.

6.3.2.2. Effect of Magnitude of Applied Pressure

Figures 6.23(a)–6.23(d) shows that a higher loading requires a longer duration before pore pressure begins to dissipate. This was consistent with the longer duration required to reach convergence point in the settlement analysis.

The void ratios at commencement of pore pressure dissipation for samples of the same initial moisture contents but under different loadings

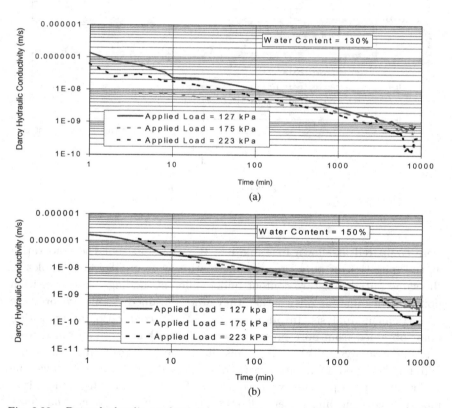

Fig. 6.20. Darcy hydraulic conductivity versus time for samples of the same moisture content but under various loadings. (a) $W = 130\%$, (b) $W = 150\%$, (c) $W = 170\%$, (d) $W = 190\%$.

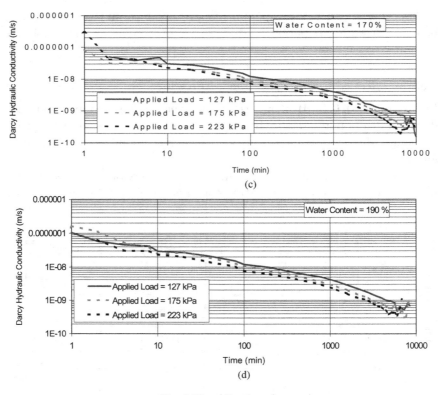

Fig. 6.20. (*Continued*).

were compared as shown in Table 6.6. It was found that the void ratios at the initiation of pore pressure dissipation were similar for samples of the same initial moisture content under different loadings. The transition points were different on whether they were obtained from pore pressure or settlement analysis. In addition to that the void ratios at transition point obtained from pore pressures were higher than those obtained from settlement analysis (Table 6.7). This difference in transition point values will be discussed in a later section. Figures 6.24(a)–6.24(d) shows the variation of pore pressure with void ratio for samples of the same initial moisture content but under different loadings. It was found that pore pressure dissipation commenced earlier when the sample was subjected to a higher loading.

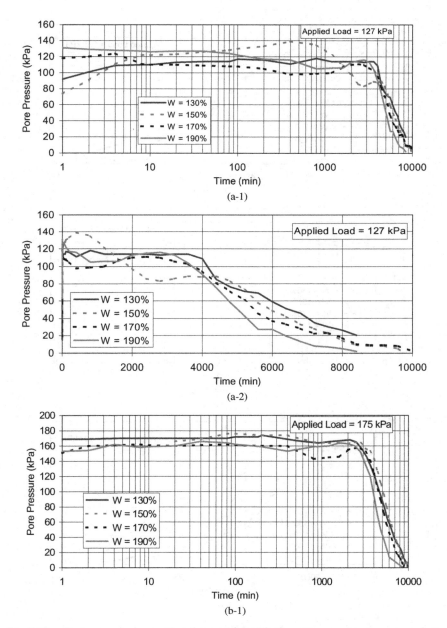

Fig. 6.21. Pore pressure versus time for samples with the various moisture contents but under the same loading. (a) 127 kPa, (b) 175 kPa, (c) 223 kPa. *Note: 1 and 2 refer to semi-log scale and arithmetic scale, respectively.

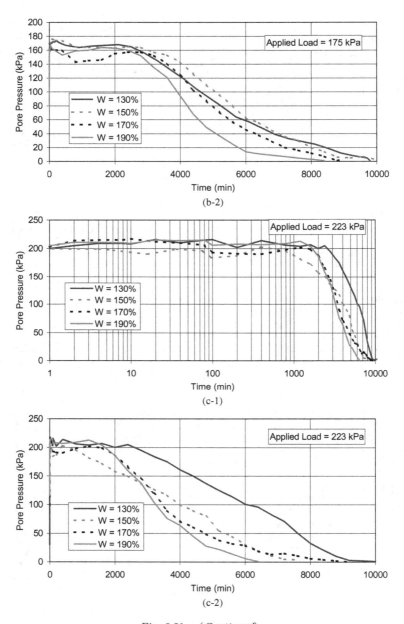

Fig. 6.21. (*Continued*).

Table 6.5. Comparison of transition point determined from settlement and time when pore pressure dissipation commences.

Moisture content (%)	Applied pressure (kPa)					
	127		175		223	
	t_s	t_p	t_s	t_p	t_s	t_p
130	4217	3727	3401	2202	2942	2350
150	4217	4541	3401	2865	2942	2512
170	4217	2512	3401	2799	2942	2078
190	4217	2955	3401	2511	2942	2022

t_s = Time required to reach transition point in settlement measurement.
t_p = Time when pore pressure dissipation commenced.

6.3.3. *Comparison of Pore Pressure from Transducers Located at Various Locations*

As expected, dissipation of pore pressure started from the drainage boundary and proceeded to the elements away from the boundary. This phenomenon was initially explained by Terzaghi and proven by Imai (1995) and Mesri and Choi (1985) with inter-connected consolidation tests. As such, comparisons were made on pore pressures measured at three locations is shown in Figs. 6.25(a)–6.25(e). Results show that the pore pressure transducer closest to the drainage boundary registered the earlier commencement of dissipation than those located further away. However, it did not occur until 1000 min after loading. At the time 30–40 mm (equivalent to 20%–27% strain) of settlement had already occurred. The delayed pore pressure dissipation at the no drain boundary for 3–6 h in the 500 mm thick sample of natural clay was explained with a family of excess pore pressure isochrones by Mesri and Choi (1985), Mesri *et al.* (1995), and Hight *et al.* (1987) based on laboratory tests. However, the delay of more than 17 h for the less than 60 mm distance from the drainage boundary indicated that the phenomenon observed in the present series of tests was different from the natural soil deformation.

6.4. **Compressibility and Consolidation Parameters in the Soil Stage**

At the onset of pore pressure dissipation, the ultra-soft slurry-like soil has transformed into a Terzaghi soil. The coefficient of consolidations c_v can then be readily determined from t_{90}. The computed average c_v values

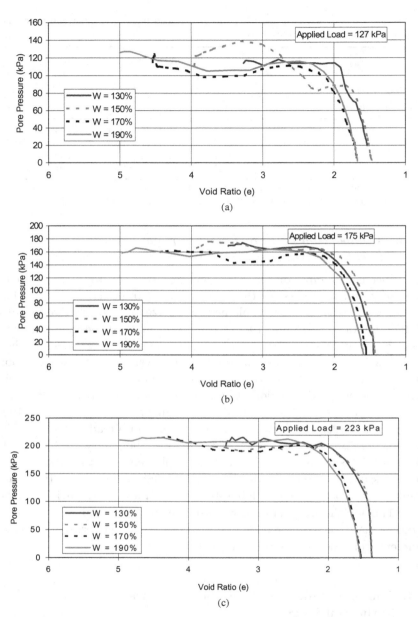

Fig. 6.22. Pore pressure versus void ratio for sample of various moisture contents under the same loading. (a) 127 kPa, (b) 175 kPa, (c) 223 kPa.

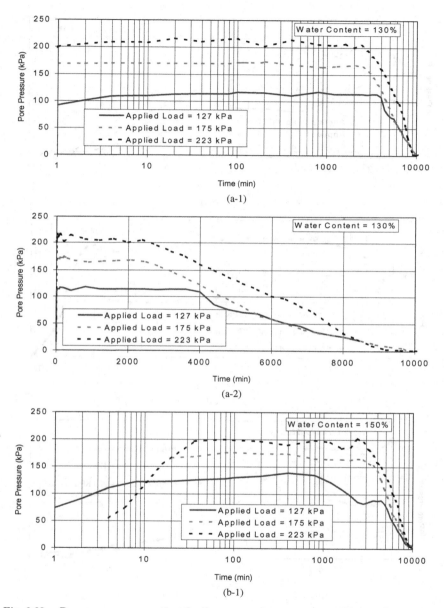

Fig. 6.23. Pore pressure versus time for the same moisture content soil but under various loadings. (a) $W = 130\%$, (b) $W = 150\%$, (c) $W = 170\%$, (d) $W = 190\%$. *Note: 1 and 2 refer to semi-log scale, and arithmetic scale, respectively.

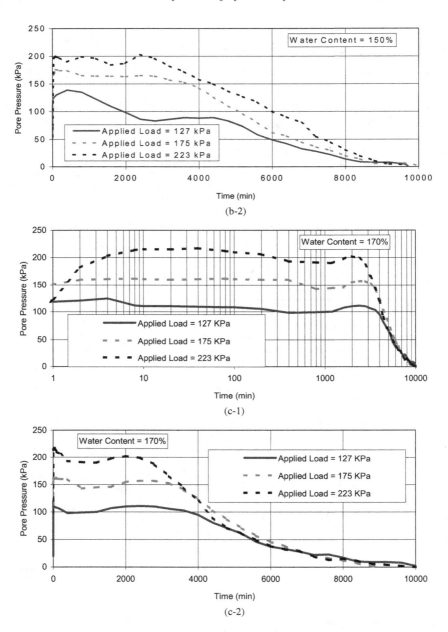

Fig. 6.23. (*Continued*).

Compression Test on Slurry with Small-Scale Consolidometer

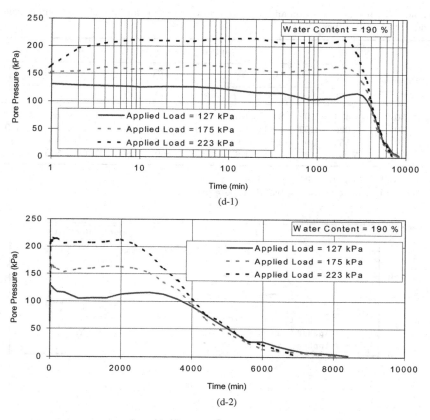

Fig. 6.23. (*Continued*).

Table 6.6. Comparison of void ratio at transition points determined from commencement of pore pressure dissipation.

Moisture content (%)	Applied pressure (kPa)					
	127		175		223	
	Time (min)	Void ratio	Time (min)	Void ratio	Time (min)	Void ratio
130	3727	2.0	2539	2.3	2350	2.13
150	4541	1.833	2865	2.13	2512	2.1
170	2512	2.6	2799	2.3	2078	2.36
190	2955	2.43	2511	2.36	2022	2.54

Table 6.7. Initial, transition and final void ratio of various initial moisture content soils under various loadings.

Moisture content (%)	e_i	Applied pressure (kPa)								
		127			175			223		
		e_{ts}	e_{tp}	e_f	e_{ts}	e_{tp}	e_f	e_{ts}	e_{tp}	e_f
130	3.484	1.92	2.0	1.535	2.0	2.3	1.450	1.9	2.13	1.374
150	4.020	1.92	1.83	1.802	2.0	2.130	1.433	1.9	2.1	1.380
170	4.556	1.92	2.6	1.683	2.0	2.300	1.555	1.9	2.360	1.537
190	5.092	1.92	2.43	1.673	2.0	2.360	1.594	1.9	2.540	1.548

e_{ts} = Void ratio at transition based on settlement.
e_{tp} = Void ratio at transition based on pore pressure.

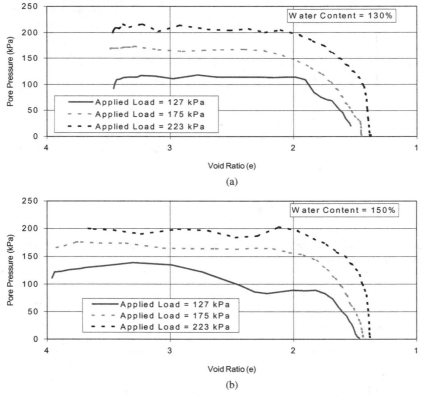

Fig. 6.24. Pore pressure versus void ratio for the same moisture content soil but under various loadings. (a) $W = 130\%$, (b) $W = 150\%$, (c) $W = 170\%$, (d) $W = 190\%$.

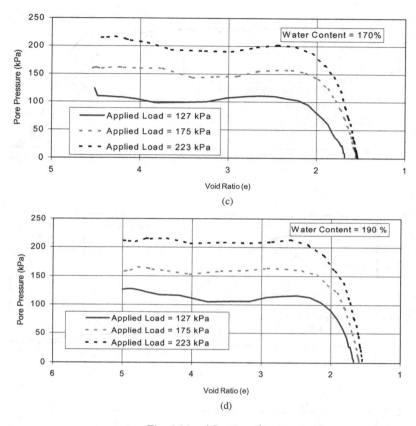

Fig. 6.24. (*Continued*).

of samples with various initial moisture contents fall between 0.72 and 0.94 m^2/yr as shown in Table 6.8.

Since the void ratio at the transition point can be determined from the settlement and pore pressure measurements, the compression index C_c can be calculated from the applied pressures and the corresponding settlements. The average void ratio change covered the period from transition point to end of primary. It is computed based on bottom piezometer measurements and using effective stress gain as the applied load. The void ratio change for a log cycle was then calculated. The computed C_c values range between 0.56 and 0.73 and are slightly higher than the intrinsic compression parameters of soils with a void ratio of about 2 (Burland, 1990).

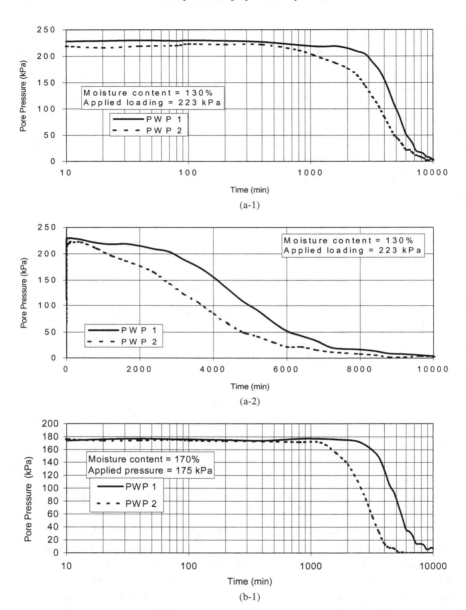

Fig. 6.25. Pore pressure versus time measured from pore pressure transducers at various locations. (a) $W = 130\%$, $\sigma = 175\,\text{kPa}$, (b) $W = 170\%$, $\sigma = 175\,\text{kPa}$, (c) $W = 170\%$, $\sigma = 223\,\text{kPa}$, (d) $W = 190\%$, $\sigma = 127\,\text{kPa}$, (e) $W = 190\%$, $\sigma = 223\,\text{kPa}$. *Note: 1 and 2 refer to semi-log scale and arithmetic scale, respectively.

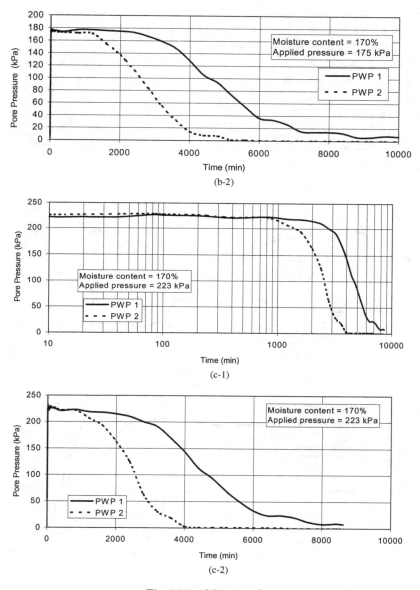

Fig. 6.25. (*Continued*).

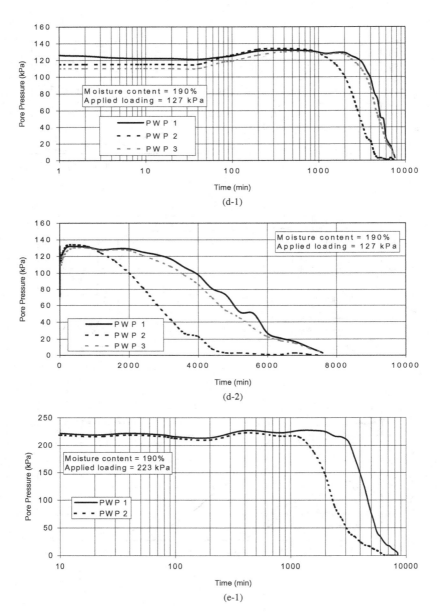

Fig. 6.25. (*Continued*).

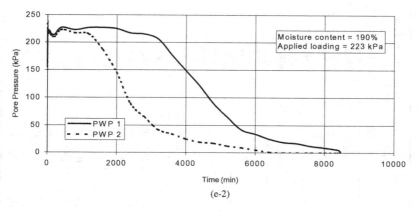

Fig. 6.25. (*Continued*).

6.5. Discussion on Void Ratio at Transition Point

The void ratio at the transition point determined based on settlement considerations was smaller than that determined from the pore pressure measurements at the bottom transducers. The possible reasons for the difference are:

(a) Transition points from settlement analysis were obtained from the change of settlement rate and convergence of void ratio values whereas transition void ratio determined from pore pressures were average void ratios at the time when the dissipation of pore pressure commenced at the bottom transducer.
(b) In reality, settlement rate of samples with different thicknesses cannot be the same even if the soil became a Terzaghi's soil. For the same soil, a thin sample settles faster than a thick sample. This point has been raised by several researchers (Aboshi, 1973; Ladd *et al.*, 1977; Tang and Imai, 1995). At the time ultra-soft soil becomes a Terzaghi's soil, the soil thicknesses were slightly different. The higher the initial moisture content the smaller the thickness of soil sample as shown in Table 6.8. Therefore, the actual void ratio at the transition must be larger than that determined from the settlement rate.
(c) If the transition void ratio was determined from the dissipation of pore pressures, different values were obtained from different initial moisture contents under the same load. It should be noted again that these void ratios were the average void ratios at the time the bottom transducer began to register dissipation of pore pressure. For a sample

Table 6.8. Comparison of compressibility parameters at soil state.

Test no.	Wi (%)	ΔP (kPa)	Initial height of the sample (mm)	Elapsed time when pore pressure dissipation (t_p) (min)	Settlement at (t_p) (min)	Drainage path (mm)	Time required for 90% consolidation (t_{90}) (min)
13–13	130	127	150	3727	50	100	5063
13–17	130	175	150	2202	37	113	6771
13–22	130	223	150	2350	47	103	6573
15–13	150	127	150	4541	66	84	5870
15–17	150	175	150	2865	55	95	5472
15–22	150	223	150	2512	57	98	5761
17–13	170	127	150	2512	53	97	5761
17–17	170	175	150	2799	62	88	4826
17–22	170	223	150	2078	62	88	5118
19–13	190	127	150	2955	67	83	3857
19–17	190	175	150	2511	67	86	3268
19–22	190	223	150	2022	62	88	4000

Test no.	Wi (%)	ΔP (kPa)	Initial height of the sample (mm)	c_v (m²/yr)	k_v (m/s)	m_v (m²/kN)	C_c
13–13	130	127	150	0.88	$5 \times 10^{-10} - 1 \times 10^{-9}$	1.22×10^{-3}	0.42
13–17	130	175	150	0.84	$2 \times 10^{-10} - 2 \times 10^{-9}$	1.47×10^{-3}	0.68
13–22	130	223	150	0.72	$2 \times 10^{-10} - 1 \times 10^{-9}$	1.08×10^{-3}	0.56
15–13	150	127	150	0.54	$1 \times 10^{-10} - 2 \times 10^{-9}$	7.79×10^{-5}	0.025
15–17	150	175	150	0.74	$3 \times 10^{-10} - 1 \times 10^{-9}$	1.27×10^{-3}	0.56
15–22	150	223	150	0.74	$5 \times 10^{-10} - 1 \times 10^{-9}$	1.04×10^{-3}	0.53
17–13	170	127	150	0.73	$2 \times 10^{-10} - 9 \times 10^{-8}$	2.0×10^{-3}	0.83
17–17	170	175	150	0.72	$3 \times 10^{-10} - 1 \times 10^{-9}$	1.29×10^{-3}	0.63
17–22	170	223	150	0.67	$2 \times 10^{-10} - 1 \times 10^{-9}$	1.09×10^{-3}	0.56
19–13	190	127	150	0.80	$5 \times 10^{-10} - 9 \times 10^{-8}$	1.74×10^{-3}	0.68
19–17	190	175	150	0.94	$3 \times 10^{-10} - 1 \times 10^{-9}$	1.3×10^{-3}	0.62
19–22	190	223	150	0.86	$4 \times 10^{-10} - 1 \times 10^{-9}$	1.26×10^{-3}	0.73

with higher initial moisture content, the sample thickness was thinner when it became a Terzaghi soil and the corresponding drainage path was shorter. Hence, commencement of dissipation was much sooner than that of a specimen with a lower initial moisture content. Therefore, the average transition void ratio was different for samples with different moisture content.

(d) The liquid limit was the water content of a soil going from a liquid stage to a plastic stage. The void ratio at the liquid limit should correspond to the void ratio at the transition point. Wood and Wroth (1976) report that at the liquid limit a slowly drying clay slurry first begins to show a small but definite shear strength of about $1.7\,\text{kN/m}^2$. The transition void ratios interpreted from the commencement of pore pressure dissipation were comparable to those at the liquid limit ($e_L = 2.36$) as shown in Table 6.6. However, those measured from settlement analysis were significantly lower than the void ratio at the liquid limit due to the reasons explained earlier. Hight et al. (1987) reported that at the time of suspension, coefficient of earth pressure at rest (K_0) should be unity. Therefore, the transition of K_0 from 1 to $(1 - \sin \emptyset')$ may correspond to the transition phase between slurry to soil. However, due to the difficulty in measurement of σ'_h, no attempt was made to determine the transition point by such measurements.

6.6. Verification of Non-Homogeneity During Slurry Compression

The compression of a clay slurry under an applied pressure is non-uniform. The soil near the drainage boundary was more compressed and hence had a lower void ratio than the soil elements located further away from the drain. Therefore, three tests were carried out under the same magnitude of load and stopped at three different times based on the commencement of pore pressure dissipation at three pore pressure transducers. The samples from each test were then divided so that the void ratio at different locations can be determined. In Test 1, which was stopped based on the pore pressure transducer close to the drainage, the measured moisture contents varied between 71% and 115% and the void ratios varied between 1.983 and 3.145 with an average of 2.564. The void ratio at the liquid limit was 2.36. Therefore, the soil element away from the drain is still in slurry stage. In Test 2, which was stopped based on the middle pore pressure transducer, the moisture contents varied between 63% and 97% and the void ratios varied between 1.7 and 2.587 with an average of 2.144. In Test 3, which was stopped based on the pore pressure transducer at the base, the moisture contents varied between 60% and 83% and the void ratios varied between 1.62 and 2.22 with an average of 1.92. The void ratio at the liquid limit for that sample was 2.28. Therefore, it is obvious that the void ratio profile along the sample varied during the slurry deformation. At the time when the pore pressure started to dissipate at the base, the portion closer to the drainage was already beyond the transition point between the slurry and the soil.

6.7. Laboratory Tests on Compressed Samples

Physical property tests and consolidation tests were carried out on the compressed samples. It was found that moisture contents were drastically reduced. The preconsolidation pressure was slightly less than the applied pressure, which was possibly due to wall friction. Figure 6.16 shows the variation of the final void ratio with the applied pressure. However, the compression indices in the soil stage calculated from Fig. 6.16 may not be accurate because the void ratio varied with the depth along the sample and the effect of wall friction was not known. Further, tests were carried out on thin samples to minimize the errors due to variation of void ratio and wall friction. A more accurate determination of parameters shall be discussed in the next chapter based on the laboratory tests on the thin samples.

6.8. Summary

Consolidation tests on ultra-soft slurry-like soil were carried out on samples with moisture contents varying between 130% and 190% under various applied loads between 123 and 223 kPa. The small-scale consolidometer was used and it was equipped with pore pressure transducers at three locations with two at the sides and one at the bottom. The vertical displacements and pore pressure responses were monitored and analyzed.

Findings from tests on samples with different initial moisture contents but under the same applied load.

The ultra-soft soils settled similar final void ratio under the same loading regardless of the initial moisture content. The magnitude and rate of settlement were highly dependent on the initial moisture content in the slurry state when subjected to the same loading condition. The higher the initial moisture content the greater was the magnitude and rate of settlement in the viscous state. The decrease in hydraulic conductivity with time increased with initial moisture content. There was an abrupt change in the rate of settlement when the slurry transformed into a Terzaghi soil. The rate of settlement after the slurry state was independent of the initial moisture content. The rate of hydraulic conductivity reduction increased when the slurry transformed into a Terzaghi soil. In the case of settlement analysis, the time taken to reach the transition point was approximately the same for all samples with different initial moisture contents under the same loading. The hydraulic conductivity change indices increased with increasing moisture content in the slurry state. The indices became very similar in the soil state. There were little pore pressure dissipation for more than 1000 min in all cases but there were vertical displacements. The time to commencement of pore pressure dissipation at the bottom transducer was affected by sample thickness. When the pore pressure started dissipating, the dissipation rate was affected by the sample thickness. The transition void ratios determined by the various approaches were slightly different for various initial moisture contents under the same loading condition. The void ratio at the transition point determined from pore pressure consideration was larger than that determined based on settlement or hydraulic conductivity consideration.

Findings from tests on samples with the same initial moisture content but under different loadings

Samples with the same initial moisture content settled to different void ratios under different loadings. The magnitude and rate of settlement were different in the slurry state and soil state. The variation of the rate of settlement was more pronounced in the slurry state. The transition point could be determined from the rate of settlement and the commencement of pore pressure dissipation. A low applied load required a longer time to reach the transition point. The void ratio at the transition point was practically independent of the load magnitude. The rate of reduction of hydraulic conductivity with time was the same for all loading conditions. The hydraulic conductivity change indices were found to be the same for all loading conditions. The c_v values were determined from the rate of pore pressure dissipation at the transition point. The c_v values fall within a narrow range of 0.72–0.84 m²/yr, which is similar to those for most natural soft clays. The C_c values in the soil state were calculated using void ratio at the transition point. The values range from 0.56 to 0.68 and were slightly higher than the intrinsic compression index according to Burland (1990). There was a period of no or little pore pressure dissipation for more than 1000 min. The rate of pore pressure dissipation recorded at the bottom transducer was much slower than the transducer located closer to the drainage boundary. The void ratio at the transition point determined from pore pressure was higher than that from settlement analyses. In fact the transition void ratio may be even higher than that determined from pore pressure considerations.

Comparison between large-scale consolidometer and small-scale consolidometer

Both tests show delay in pore pressure dissipation during slurry stage. The small-scale consolidometer is easier to handle than the large-scale consolidometer, yet the results obtained by both are very similar. However, both tests cannot provide a meaningful e-log σ'_v relationship due to variation of void ratio along the sample. Even in the small-scale consolidometer, void ratio variation was significant as explained in Section 6.6.

Chapter 7

COMPRESSION TESTS ON ULTRA-SOFT SOIL WITH HYDRAULIC CONSOLIDATION CELL

It was found in the test with large sample thickness of 150 mm, there were substantial variations of void ratio with depth at the transition point as well as at the time when pore pressure dissipation commenced. The built up of wall friction might have affected the results in the soil stage due to low diameter-to-height ratio (D/H) of 0.6. Even in the slurry stage, silo effect would occur due to the large D/H ratio. Therefore, additional tests were carried out on 40-mm thick samples with different moisture contents using the large diameter (150 mm) hydraulic consolidation Rowe cell. The cell has a D/H ratio of 3.75 which satisfies the D/H ratio of 3 recommended by Mesri et al. (1995) and Tsukada and Yasuhara (1995). Tests were carried out using a single-step instantaneous loading and multiple-step incremental loading.

7.1. One-Step High Pressure Loading Tests with Hydraulic Consolidation Cell

7.1.1. *Description of the Apparatus*

The apparatus used is the Rowe cell, which was first developed at Manchester University (Rowe, 1966). The inner diameter of the apparatus was 150.5 mm and the sample thickness is approximately 40 mm. Therefore, the diameter-to-thickness ratio was about 3.8. During the slurry stage, no friction was expected. Although wall friction developed when the material transformed into a Terzaghi soil, the diameter-to-thickness ratio had become much higher and then wall friction still remained insignificant. The rigid loading plate was used to ensure an equal strain loading condition.

Before pouring the sample, the cell wall was coated with silicone grease to reduce the wall friction.

The soil settlement was measured using a linear vertical displacement transducer (LVDT type HS25B, Model No. WF 17015 manufactured by Wykeham Farrance, England). It had a maximum travel distance of 25 mm and an accuracy of 0.001 mm. The LVDT was attached to a piston rod. The pore pressure was measured at the base using a pore pressure transducer (Model No. WF 17021 manufactured by Wykeham, Farrance, England), which had an accuracy of 1 kPa. The load was applied through a G.D.S. hydraulic controller system manufactured by G.D.S Instrument Ltd. with an accuracy of 1 kPa. All the measured data were logged by an advanced data logger AT2000-32 manufactured by Wykeham Farrance, England. The test layout and details of the equipment is shown in Fig. 7.1. Before the test was carried out, the cell was calibrated to determine the difference between the applied pressure and the net pressure on the soil.

7.1.2. Sample Preparation and Method of Testing

Instantaneous one-step loadings were applied to the slurry-like samples with different moisture contents. The combinations of moisture content and loading are shown in Table 7.1. Samples were collected from the siltpond and prepared to the desired moisture content according to Eq. (4.2). The slurry was then poured into the cell.

After reaching the required thickness, the slurry was thoroughly mixed to remove the trapped air. A total of 12 samples were prepared with moisture contents of 130%, 150%, 168%, and 190% and subjected to three different one-step loadings of 62, 118, and 230 kPa.

Since the samples were saturated, no back-pressure was applied. During the test, the vertical displacement, pore pressure, and volume of water drained out from the samples were measured.

7.1.3. Discussion of Results

7.1.3.1. Pore Pressure Response

In all tests, the pore water pressures responded instantaneously and were almost equal to the applied pressure confirming that the sample were in a saturated condition. There was no dissipation of pore pressure for a long period of time while the settlement continued. This phenomenon was

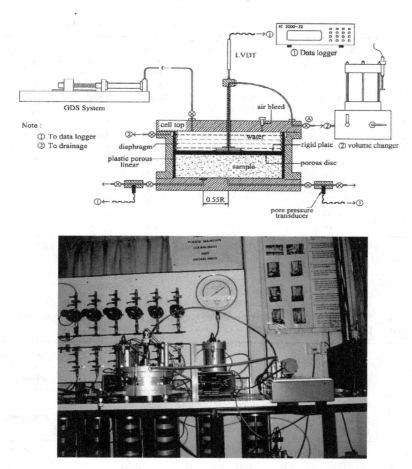

Fig. 7.1. Arrangement of hydraulic Rowe cell test.

reported in Chapters 5 and 6 and in Bo *et al.* (1999, 2001a, b). A comparison of the applied pressures and pore pressures response together with the delay in pore pressure dissipation is shown in Table 7.2.

As can be seen in Table 7.2 and Figs. 7.2(a)–7.2(d) there is a period with little or no pore pressure dissipation at the bottom of the sample. The period of delayed dissipation pore pressure can be as long as 200–400 min in most cases. This delay is longer in lower pressure loadings than at higher pressure loadings. This phenomenon was also reported in Chapter 6 and by Bo *et al.* (2001b).

Table 7.1. Initial condition of sample and applied pressure.

Sr. no.	Thickness of the sample (mm)	Diameter of the sample (mm)	Moisture content (%)	Void ratio	Applied pressure (kPa)
1	39.70	150.50	130.02	3.583	62
2	39.92	150.50	130.92	3.583	118
3	39.46	150.50	131.43	3.583	230
4	38.89	150.50	150.37	4.046	63
5	40.00	150.50	150.03	4.046	118
6	41.00	150.50	150.39	4.046	230
7	39.80	150.50	168.04	4.513	63
8	40.00	150.50	168.29	4.513	118
9	39.59	150.50	168.44	4.513	230
10	39.53	150.50	190.21	5.151	62
11	41.53	150.50	190.26	5.151	118
12	38.19	150.50	190.20	5.151	230

Table 7.2. Comparison of applied pressure, responded pore pressure and duration of delayed in pore pressure dissipation.

Test code	Applied pressure (kPa)	Initial pore pressure (kPa)	Duration of delayed in pore pressure t_i (min)	Settlement at t_i time		Strain[a] (%)	Remaining thickness when pore pressure dissipation commence (mm)
				From volume change (mm)	From LVDT (mm)		
50/130	62.00	58.60	400	11.50	10.35	29.00	29.35
100/130	118.00	119.20	(200)[b]	10.62	10.53	26.38	29.39
200/130	230.00	228.20	(200)[b]	13.65	12.46	31.58	27.00
50/150	63.00	61.80	(300)[b]	12.00	13.00	30.00	25.00
100/150	118.00	116.20	(200)[b]	14.37	12.04	30.10	28.00
200/150	230.00	231.40	(200)[b]	19.50	18.00	43.80	23.00
50/168	63.00	63.00	(400)[b]	15.70	14.30	36.00	25.50
100/168	118.00	118.00	(200)[b]	15.70	14.42	27.74	25.58
200/168	230.00	230.60	(200)[b]	17.52	16.29	41.15	23.30
50/190	62.00	62.00	(200)[b]	14.56	13.00	32.90	26.53
100/190	118.00	118.60	(200)[b]	17.45	17.26	41.56	24.27
200/190	230.00	231.20	(200)[b]	20.45	18.78	49.17	19.41

[a] From settlement.
[b] Little pore pressure dissipation.

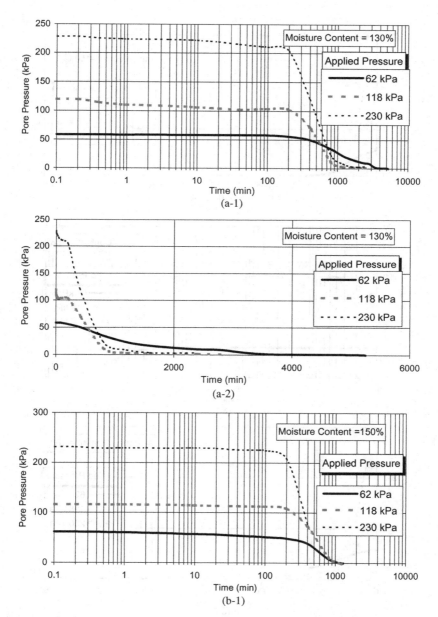

Fig. 7.2. Pore pressure versus time for samples with the same moisture content but under various loadings W = (a) 130%, (b) 150%, (c) 168%, (d) 190%. (1 and 2 refer to semi-log scale and arithmetic scale, respectively.)

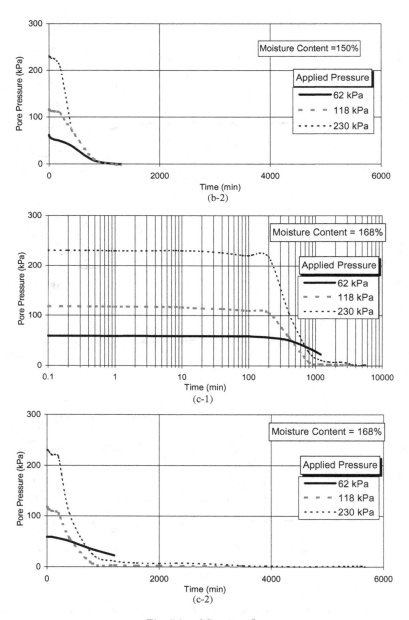

Fig. 7.2. (*Continued*).

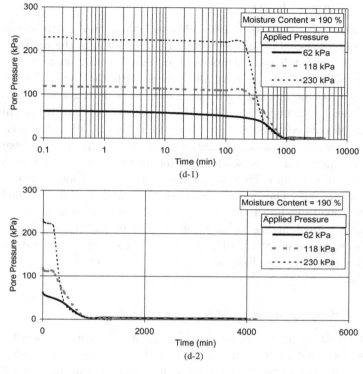

Fig. 7.2. (*Continued*).

At the time when pore pressure dissipation commenced, the sample thicknesses were found to range between 19 and 29 mm only. The higher the initial water content the smaller is the thickness of the remaining sample. For samples with the same moisture content but subjected to various loads, the higher the load magnitude, the smaller is the thickness of the remaining sample at the time when pore pressure dissipation commenced at the base. Furthermore, the rate of pore pressure dissipation varies with the load magnitude. The higher the applied load, the greater is the rate of dissipation due to a greater hydraulic gradient and shorter drainage path. For samples with different moisture contents but under the same load, the rate of dissipation in the soil stage is again different. The higher the initial moisture content, the greater is the pore pressure dissipation rate due to shorter drainage of the thinner samples. The vertical displacement during this stage would be discussed in the following section.

7.1.3.2. Settlement Response

As reported by Bo et al. (1999), although there were little or no pore pressure dissipation in the early stage of compression in the ultra-soft soil, significant vertical displacements were measured in that stage.

The same behavior was found in tests with the hydraulic cell. The measured settlements tallied with the settlements determined from volume change measurements (Fig. 7.3). The measured settlements during the delayed dissipation period are also shown in Table 7.2. It is obvious that the magnitude of settlement or volumetric strain during the delayed dissipation period is significant. The settlement versus time for samples of various moisture contents subjected to various loading conditions are shown in Figs. 7.4 and 7.5. The increase in settlement with initial void ratio and additional load during this early stage is consistent with findings reported in Chapter 6 involving the small-scale consolidometer tests and by Bo et al. (2001a). It can be noted that the percentage of strain occurred during the delayed of pore pressure dissipation period was found to range between 26% and 49%.

7.1.3.3. Rate of Settlement and Hydraulic Conductivity

To find out the compression behavior during the slurry stage, the settlement rates were plotted against time as shown in Figs. 7.6 and 7.7. It can be seen in these figures that the larger the initial moisture content, the greater is the settlement rate. Furthermore, a greater settlement rate was noted when the

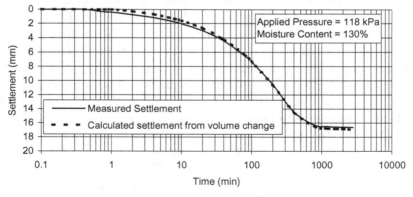

Fig. 7.3. Comparison of settlements measured from LVDT and calculated from volume change.

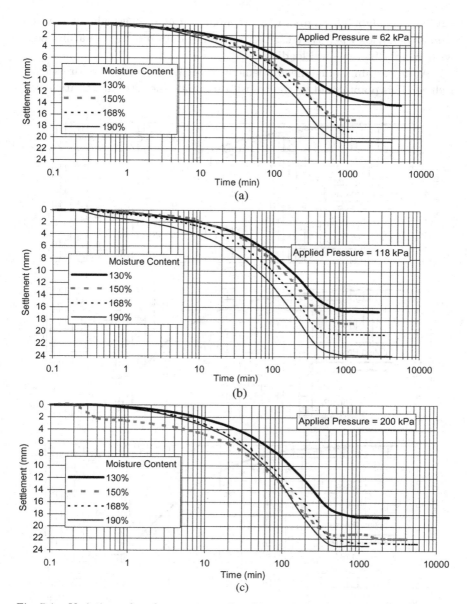

Fig. 7.4. Variations of settlement versus time for various moisture contents but under the same loading, Applied pressure = (a) 62 kPa, (b) 118 kPa, (c) 230 kPa.

loading was high. This agrees with Yoshikuni *et al.* (1995a) who observed a similar behavior when they carried out consolidation tests on natural soils with various incremental loading.

At a certain point the settlement rate was observed to drastically reduce. This point could be the transition point. The settlement rate at the

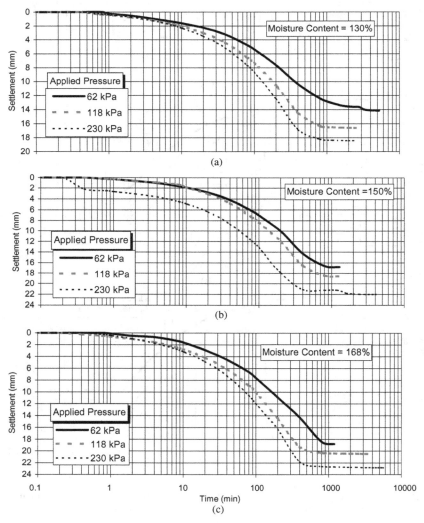

Fig. 7.5. Variations of settlement versus time for the same moisture content but under various loadings W = (a) 130%, (b) 150%, (c) 168%, (d) 190%.

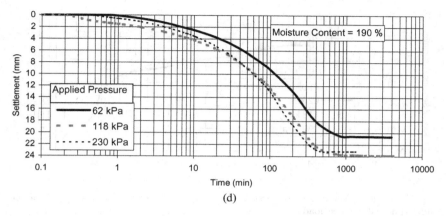

Fig. 7.5. (*Continued*).

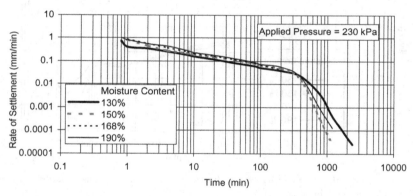

Fig. 7.6. Rate of settlement versus time for samples of various moisture contents but under the same magnitude of loading.

transition point ranges between 0.01 mm/min and 0.08 mm/min. The higher the moisture content and additional load, the greater is the settlement rate at the transition point. This finding was consistent with the change of settlement rate at the transition point reported in the earlier chapter. Since the settlement rate is directly related to the hydraulic conductivity in the viscous stage, the average hydraulic conductivity is calculated from the volume of discharge using the following equation:

$$k = \frac{Q}{Ai} \tag{7.1}$$

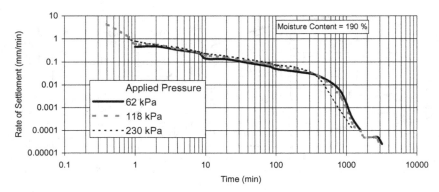

Fig. 7.7. Rate of settlement versus time for samples of the same moisture content but under different applied loads.

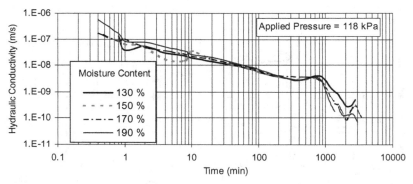

Fig. 7.8. Hydraulic conductivity versus time for samples with various moisture contents but under the same loading.

where Q is the volume of water discharged in m^3/s, A is the area of sample in m^2, $i = h/l$ is the hydraulic gradient and is dimensionless, h is the head of water pressure applied in metre and l is the height of sample in metre. As shown in Fig. 7.8 the hydraulic conductivity decreases at a constant rate in the slurry stage. When the material turned into a Terzaghi soil, a sudden decrease in hydraulic conductivity was noted. Furthermore, the hydraulic conductivity decreases with increasing applied pressure (Fig. 7.9). For finite strain, the relationship between void ratio, hydraulic conductivity and additional load are important (Van Essen et al., 1995). The relationship between the hydraulic conductivity and the void ratio is shown in Figs. 7.10 and 7.11. The hydraulic conductivity reduction

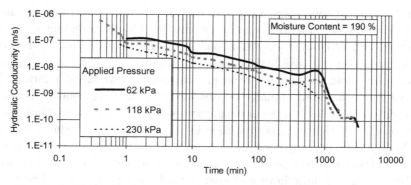

Fig. 7.9. Hydraulic conductivity versus time for samples of the same moisture content but under various loadings.

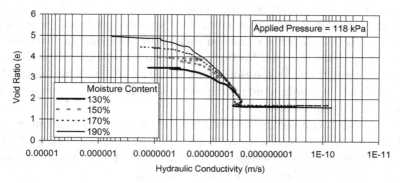

Fig. 7.10. Void ratio versus hydraulic conductivity for samples of various moisture contents under the same loading.

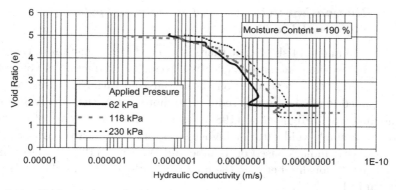

Fig. 7.11. Void ratio versus hydraulic conductivity for the same moisture content under various loadings.

ratio in the viscous stage ranges from 1.53 to 3.19. The ratio becomes higher with increasing initial moisture content. This is consistent with the findings by Yoshikuni et al. (1995b) on the void ratio–hydraulic conductivity relationship for soil with high moisture content. However, it can be seen in the figures that the void ratio versus hydraulic conductivity relationship is not linear and can actually be divided into three segments. The non-linearity of the e-log k curve in the slurry soil is also reported by Monte and Krizek (1976). However, as explained in the earlier chapter, the reduction of hydraulic conductivity with minimum change in void ratio in the early stage of test could be due to the formation of mud cake during that period. When fine grains may have migrated toward the porous stone. The same hydraulic conductivity measured for varied void ratio in Fig. 7.10 may also have been partly caused by mud cake formation. The slight difference in interpreted hydraulic conductivity for the same void ratio in Fig. 7.12 is also possible due to the assumption of mean pore pressure. It was found that the hydraulic conductivity reduction ratios were very similar to the values reported in Chapter 6 and by Bo et al. (2001b) based on the small-scale consolidometer tests. At the time when the material turns into a soil, the corresponding hydraulic conductivity is very similar to the values normally found in the natural clays.

The e-log k relationship obtained from small-scale consolidometer tests and Rowe cell tests were compared in Fig. 7.12. The results are similar. Slight differences could be due to simplification of pore pressure profile assumed in the calculation of hydraulic conductivities.

7.1.4. *Relationship Between Void Ratio and Applied Additional Load*

Figures 7.13 and 7.14 show that there is a very large reduction in the void ratio with very little increase in effective stress. The reduction of void ratio can be as high as 3 depending on the initial void ratio. However, for the same applied load, the sample reduced to more or less the same void ratio at the final stage. Hight et al. (1987) described the convergence of e-log σ'_v curves to a single void ratio at higher stresses for reconstituted clay with various plastic indices based on Lambe and Whitman (1969) experimental data. The relationship between the final void ratio and the applied pressure is shown in Fig. 7.15.

It can be seen that the higher the applied load, the smaller is the final void ratio. Tan et al. (1988) have also carried out compression tests on

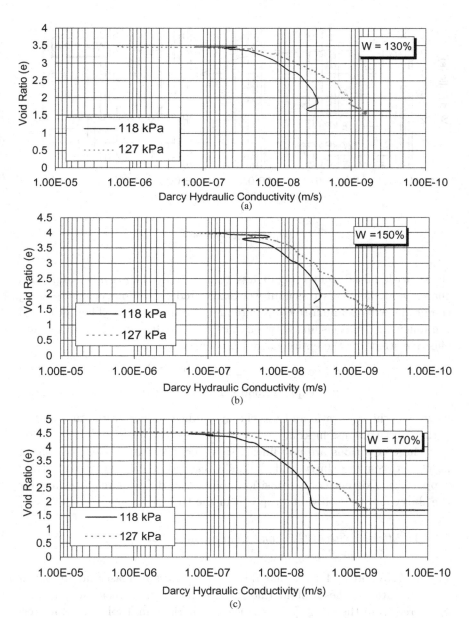

Fig. 7.12. Comparison of e-$\log k$ relationship between two type of tests under similar loadings. (a) $W = 130\%$, (b) $W = 150\%$, (c) $W = 170\%$, (d) $W = 190\%$

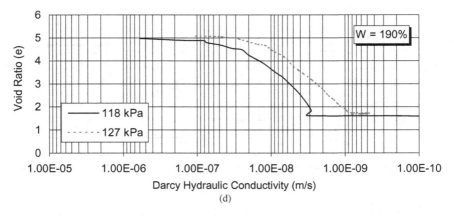

Fig. 7.12. (*Continued*).

slurry with initial void ratio of less than 6.7. In their study, void ratio change and effective stress gain were measured for various elements along the profile with Gamma-ray and pore pressure transducers. They proposed the following relationships between void ratio, hydraulic conductivity, and effective stress for the entire range from slurry to soil:

$$\sigma' = \exp(6.887 - 2.205e + 0.157e^2) \quad (7.2)$$
$$k = \exp(-22.366 + 1.468e - 0.090e^2) \quad (7.3)$$

where e is the void ratio and k is the hydraulic conductivity in m/s and σ' is the effective stress in kPa.

These equations are non-linear. However, there is no clear transition point between the slurry and the soil. Test results of the present study support the existence of a transition point. Two different relationships for the slurry and the soil were proposed in the later section.

7.1.4.1. *Effective Stress Gain and Compression Index (C_{c1}^*) in Viscous Stage*

Figures 7.13 and 7.14 show that the effective stress increases rapidly when the void ratio reaches around 1.90–2.4. Details of the transition void ratio measured from the e-log σ'_v curves are summarized in Table 7.3. However, in this method the transition point is taken as the interception between the two curves. The compression index for the slurry stage can be obtained from the change in void ratio in the log cycle from 1 to 10 kPa (Fig. 7.15).

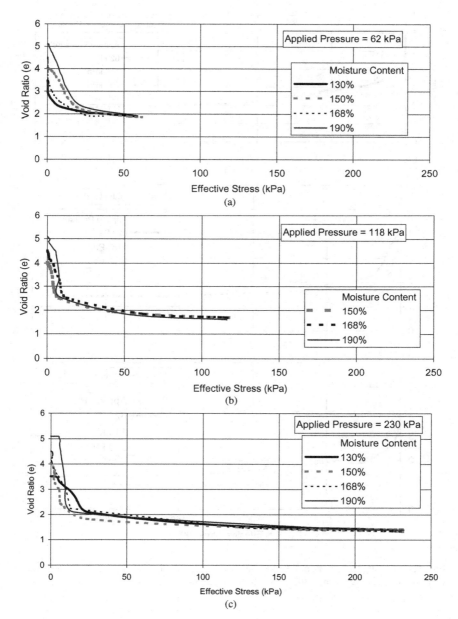

Fig. 7.13. Void ratio versus effective stress for various moisture contents under the same loading. Applied pressure = (a) 62 kPa, (b) 118 kPa, (c) 230 kPa.

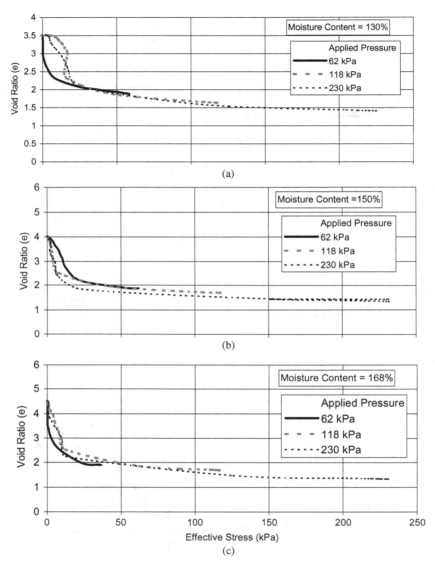

Fig. 7.14. Void ratio versus effective stress for the same moisture content soil under various loadings. $W =$ (a) 130%, (b) 150%, (c) 168%, (d) 190%.

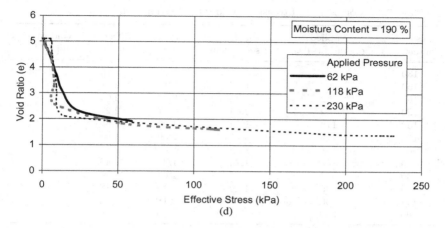

Fig. 7.14. (*Continued*).

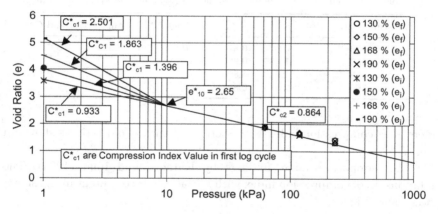

Fig. 7.15. C_{c1}^* determined assuming intercept at e_{10}^*.

The values obtained are summarized in Table 7.4. The void ratio change before 1 kPa can be considered negligible.

The results are consistent with high compression indices below 10 kPa reported by Kim *et al.* (1995) for high moisture content soil. Lee and Sills (1981) also stated that when a surface loading is applied to soft soil, the majority of the compression occurred when the effective stresses reach equilibrium with the soil weight. The compression indices in the slurry stage were correlated with the initial void ratios. There is a clear trend of

Table 7.3. Comparison of transition void ratio determined by various methods.

Moisture content (%)	Applied pressure (kPa)	Change in rate of settlement	Pore pressure	Effective stress	Void ratio at liquid limit
130	62	1.95/2.09	2.21	2.16	2–2.3
150	62	1.88	2.17	2.17	
168	62	1.92	2.69	2.63	
190	62	2.00	2.27	2.27	
130	118	1.70	1.93	1.88	
150	118	1.75	2.00	1.98	
168	118	1.80	2.432	2.45	
190	118	1.70	2.4/1.86	2.49	
130	230	1.80	2.04	2.00	
150	230	1.80	1.79	1.88	
168	230	—	2.18	2.12	
190	230	—	2.00	2.00	

Table 7.4. Comparison of compression indices (assuming interception at e_{10}).

Moisture content (%)	C_{c1}^*	C_{c2}^*	Average C_{c2}^*	e_{10}	Burland C_c^*
130	0.977	0.825	0.846	2.650	0.535
150	1.455	0.930			
168	1.909	0.890			
190	2.545	0.841			

increasing compression index with increasing initial void ratio as shown in Fig. 7.16.

This supports similar findings by Katagiri and Imai (1994). The relationship of compression index with initial moisture content in the slurry stage is as follows:

$$C_{c1}^* = e_i - e_{10}^* \tag{7.4}$$

where C_{c1}^* is the compression index in viscous stage, e_i is initial void ratio, and e_{10}^* is the void ratio at 10 kPa effective stress and usually found to be about 2.65 for the tests.

It is assumed that the lines defining the indices at the viscous and soil stages intersect at the stress of 10 kPa. Therefore, e_{10}^* becomes an important parameter for the determination of compression indices. It can be related with the void ratio at the liquid limit as follows:

$$e_{10}^* = \alpha e_L \tag{7.5}$$

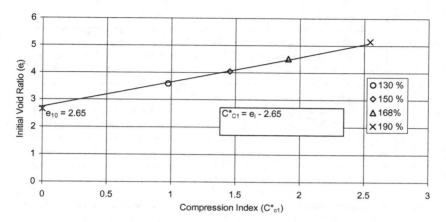

Fig. 7.16. Initial void ratio versus compression indices in first log cycle (C_{c1}^*).

where e_L is the void ratio at liquid limit, α is a constant which range between 0.9 and 1.32. This correlation will be discussed in a later section.

The compression index C_{c2}^* in the soil stage (i.e., between 10 kPa and 100 kPa) falls within a narrow range of 0.825–0.93 with a mean value of 0.864. C_{c2}^* is slightly higher than Burland (1990) intrinsic compression index C_c^* which measured between e_{100} and e_{1000}. These compression parameters will also be confirmed in a later section. If C_{c1}^* were worked out based on the intercept of the slopes at the transition void ratio e_t determined from the commencement of pore pressure dissipation, C_{c1}^* values were found to decrease with increasing applied pressure (Fig. 7.17). The relationship of

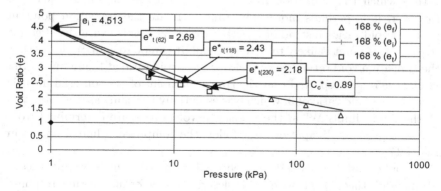

Fig. 7.17. C_{c1}^* determined assuming intercept of first slope at transition void ratio.

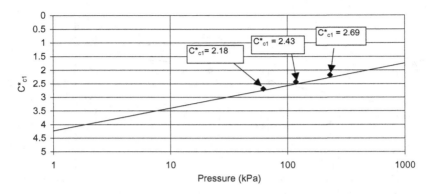

Fig. 7.18. C^*_{c1} determined from intercept at transition void ratio versus pressure.

C^*_{c1} with applied pressure for this case with 168% moisture content soil is shown in Fig. 7.18.

7.1.4.2. Transition Void Ratio

For settlement analysis involving an overconsolidated soil, the important parameters are the compression indices and preconsolidation pressure. Likewise in the prediction of settlement involving an ultra-soft soil, the parameters required are the compression index in the slurry stage and the transition point in which a slurry-like soil transformed into a Terzarghi soil.

Monte and Krizek (1976) stated that there is a critical stress range within which the transition from a slurry to a normal soil takes place. However, it is perhaps more important to relate the transition point in terms of void ratios instead of effective stress. This is because the effective stress at this point is likely to be very small. It can be seen in Table 7.3 that the transition void ratios are very close to the void ratio at liquid limit. Kim et al. (1995) also reported that the transition point of the soil used in their study had a void ratio of 3, which was essentially the same as the void ratio at the liquid limit. When the transition void ratios are determined from the pore pressure dissipation they follow the compression line as shown in Fig. 7.17.

Monte and Krizek (1976) reported that the small strain theory (Terzarghi's theory) can adequately describe the deformation-time response for all practical purposes after the effective stress exceeds about 55 kPa.

7.2. Step Loading Compression Tests on Ultra-Soft Soil with Hydraulic Consolidation Cell

To better understand the compressibility of this ultra-soft soil, further step loading tests were carried out with a hydraulic consolidation cell with various drainage conditions.

7.2.1. Method of Tests

The Rowe cell (WF 24557) has an inside diameter of 150.5 mm. Three different drainage conditions were adopted. They were radial outward, radial inward, and both inward and outward. The porous liner used in the outward drainage case was a plastic liner with a hydraulic conductivity of 4×10^{-4} m/s. For the inward drainage case, the central drain was 7.5 mm diameter and was made of highly permeable geotexile with cross plane hydraulic conductivity of 15×10^{-4} m/s. Before the test, the Rowe cell diaphragm is also calibrated to determine the net pressure applied on the sample.

All samples were about 40 mm thick. For the inward radial drainage case, the sample diameter was 150.5 mm. For the other two drainage conditions, the diameter was 147 mm (Table 7.5).

The samples were collected from the waste pond submerged under 7 m of seawater in the foreshore area at the eastern part of Singapore. Before pouring the sample into the Rowe cell, it was thoroughly stirred to remove any entrapped air. The desired moisture content was prepared using the procedure described in Chapter 4.

Tests were carried out on the ultra-soft soils at two different moisture contents of 150% and 168%. Tests were carried out with multi-step load

Table 7.5. Details of tests.

Test code	Drainage	Sample diameter (mm)	Drainage size filter (mm)	Thickness of sample (mm)
O	Radial outward	147	1.75 mm thickness at side	40
I	Radial inward	150.5	7.50 mm diameter in the center	40
D	Both radial inward and outward	147	1.75 mm thickness at side and 7.5 mm diameter in the center	40

Table 7.6. Details of loadings.

Test code	Moisture content (%)	Void ratio	Initial load (kPa)	Final load (kPa)	load increment ratio	Drainage condition	Type of test
150 O	150.06	4.046	12	400	1	Radial outward	Equal strain
150 I	150.07	4.046	12	400	1	Radial inward	Equal strain
150 D	150	4.044	14	400	1	Radial inward & outward	Equal strain
170 O	168.18	4.513	10	400	1	Radial outward	Equal strain
170 I	168.18	4.512	12	400	1	Radial inward	Equal strain
170 D	168.18	4.513	6	640	1	Radial inward & outward	Equal strain

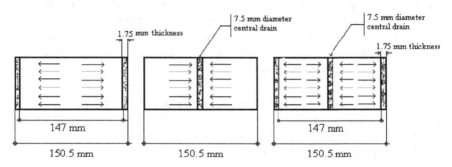

Fig. 7.19. Diagram showing arrangement of drainage in compression tests.

(Table 7.6). The first loading started with a minimum of 6 kPa and eventually increased to a maximum of 640 kPa. Tests were carried out with an equal-strain condition. Details of measurements and equipments used are described in the earlier section. Details of the test equipment are shown in Fig. 7.1. The layout of tests is similar to the layout described in earlier section. The drainage conditions are shown in Fig. 7.19.

7.2.2. Discussion of Results

7.2.2.1. Settlement Characteristics

The complete settlement records of all six tests are shown in Figs. 7.20(a) and 7.20(b). It was found that unlike the natural soil, the ultra-soft soil settled a very large amount in the first load step of less than 12 kPa. In the test with $W = 168\%$, the void ratio reduced from 4.5 to about

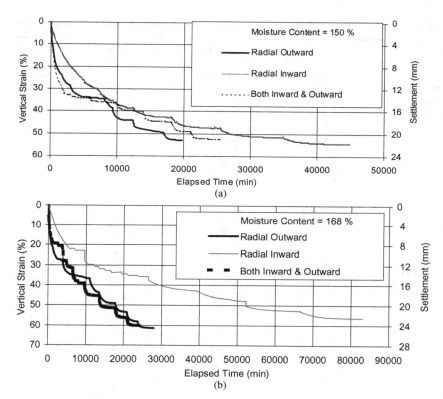

Fig. 7.20. Variations of vertical strain and settlement with time. $W =$ (a) 150%, (b) 168%.

3.0 under the first loading. For the sample with $W = 150\%$, the void ratio reduced from 4.046 to about 2.6 under the first loading. The magnitude of settlement or the change in void ratio decreases gradually with subsequent load steps as shown in Table 7.7. The e-$\log \sigma'_v$ curves for the three different drainage conditions are shown in Figs. 7.21(a) and 7.21(b). Tests for all three drainage conditions yielded almost the same results. It was noted that the inward drainage condition yielded slightly smaller settlement. It is interesting to note that these curves can be divided into three sections separated by the three log cycles. The slope or compression index of each section decreases with increasing pressure. At the third log cycle, the compression index is comparable to the intrinsic compression index of natural soil reported by Burland (1990) as shown in Table 7.8. Therefore, e-$\log \sigma'_v$ curves for ultra-soft

Table 7.7. Comparison of magnitude of settlement and change in void ratio in each step.

Test code	Loading step	Magnitude of settlement during the step (mm)	Magnitude of void ratio change (mm)	Final void ratio
150 O	12	11.201	1.413	
	25	2.071	0.261	
	50	1.647	0.208	
	100	2.192	0.277	
	200	2.260	0.285	
	400	1.614	0.204	1.399[a]
150 I	12	12.261	1.546	
	25	4.310	0.544	
	50	1.540	0.194	
	100	2.142	0.270	
	200	1.835	0.232	
	400	1.240	1.156	1.103[a]
150 D	14	11.99	1.513	
	18	1.109	0.140	
	25	1.610	0.203	
	50	1.660	0.209	
	100	2.210	0.279	
	200	1.970	0.249	
	400	1.700	0.214	
168 O	10	11.893	1.639	
	20	2.970	0.409	
	40	3.143	0.433	
	80	2.342	0.323	
	160	1.947	0.268	
	320	1.986	0.274	
	640	1.697	0.234	0.932[a]
168 I	12	12.13	1.670	
	25	4.090	0.563	
	50	2.930	0.404	
	100	2.000	0.276	
	200	1.530	0.211	
	400	1.350	0.186	1.20[a]
168 D	6	9.852	1.358	
	12	2.816	0.388	
	25	2.651	0.365	
	50	2.861	0.394	
	100	2.100	0.289	
	200	1.667	0.230	
	400	1.803	0.248	1.24[a]

[a] Measured average values.

Table 7.8. Comparison of compression parameters.

| | Moisture content = 150% | | | Moisture content = 168% | | |
| | Drainage condition | | | Drainage condition | | |
	Outward	Inward	Both in and out	Outward	Inward	Both in and out
e_i	4.046	4.046	4.046	4.513	4.513	4.513
Measured e_{10}^*	2.700	2.700	2.730	2.874	3.300	2.88
Measured e_{100}^*	1.786	1.670	1.707	1.639	1.874	1.684
Measured e_f	1.319	1.330	1.240	0.933	1.352	1.197
Calculated e_{10}	2.789					
Calculated e_{100}	1.642					
Measured C_{c1}^*	1.346	1.346	1.316	1.639	1.212	1.633
Calculated C_{c1}^*	1.256			1.723		
Measured C_{c2}^*	0.914	1.030	1.020	1.235	1.426	1.196
Calculated C_{c2}^*	1.204			1.149		
Measured C_{c3}^*	0.861	0.610	0.870	0.882	0.763	0.814
Calculated C_{c3}^*	0.660					
Measured e_L	2.760					

soil as well as reconstituted soil is not linear and may have two or three slopes in three log cycles. This type of material non-linearity including non-linear stress–strain behavior for natural clays has been discussed by Rowe (1972), Jardine et al. (1986), Jardine and Hight (1987), and Hight et al. (1987).

The e-$\log \sigma_v'$ curves obtained from the two different water contents (150% and 168%) with various radial drainage conditions are shown in Figs. 7.22(a)–7.22(c). The shapes of the two curves follow a similar trend in second and third log cycles. The higher water content, the greater is the compression indices at the first two log cycles.

7.2.2.2. Pore Pressure Characteristics

At the first step of loading, the measured pore pressure was slightly lower than the applied pressure. In all tests, there was a period with little or no pore pressure dissipation. The duration was found to be in the range between 1 and 40 min as shown in Figs. 7.23 and 7.24. At higher load increments exceeding 160 kPa, there appeared to be an increase in pore pressure probably due to the Mandel–Cryer effect (Schiffman et al. 1969) as well as a delay in the system due particularly to side friction.

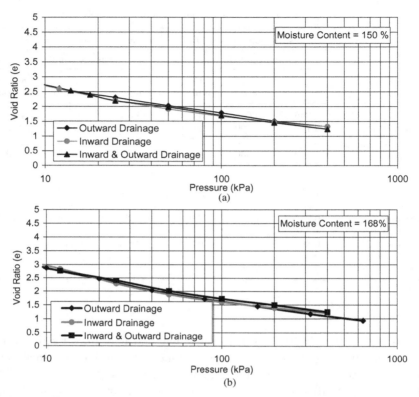

Fig. 7.21. Variations of void ratio with pressure for samples with various drainage conditions. $W =$ (a) 150%, (b) 168%.

7.2.3. Correlation of Compression Index with Physical Parameters

The correlations between index tests and the properties of remoulded clays had been reported by Carrier III and Beckman (1984). They introduced new parameters β, α, and ε which correlated well with the Atterberg limits, initial water content, and activity as follows:

$$\alpha = (0.0208)\ \mathrm{PI}[1.192 + (\mathrm{act})^{-1}] \tag{7.6}$$

$$\beta = -0.143 \tag{7.7}$$

$$\varepsilon = 0.027(\mathrm{PL}) - 0.0133(\mathrm{PI})[1.192 + (\mathrm{act})^{-1}] \tag{7.8}$$

where PL is the plastic limit, PI is the plastic index, and act is the activity.

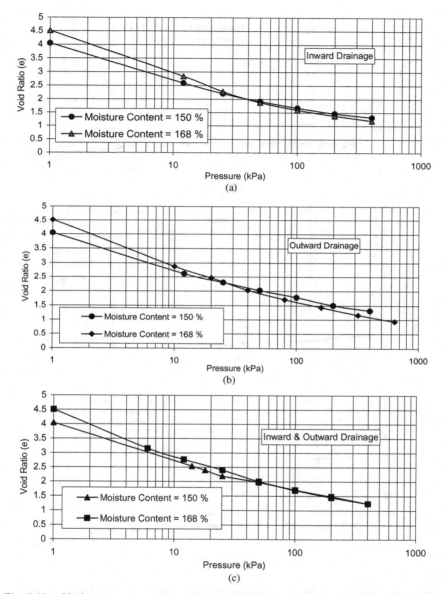

Fig. 7.22. Void ratio versus pressure from two different moisture contents. (a) Inward drainage, (b) outward drainage, (c) inward and outward drainage.

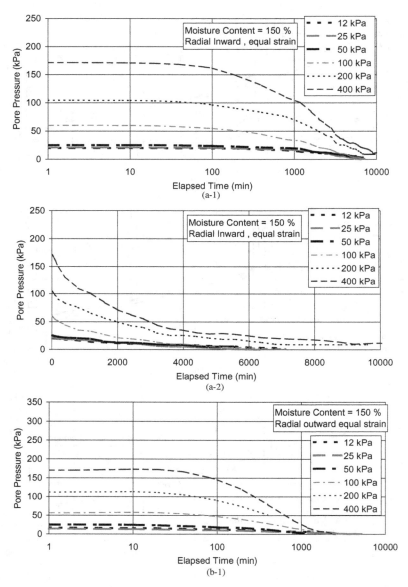

Fig. 7.23. Pore pressure versus time from various loading steps ($W = 150\%$). (a) Inward drainage, (b) outward drainage, (c) inward and outward drainage. (1 and 2 refer to semi-log scale and arithmetic scale, respectively.)

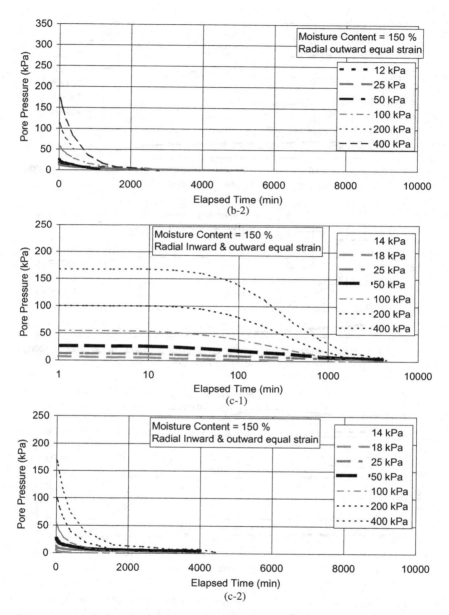

Fig. 7.23. (*Continued*).

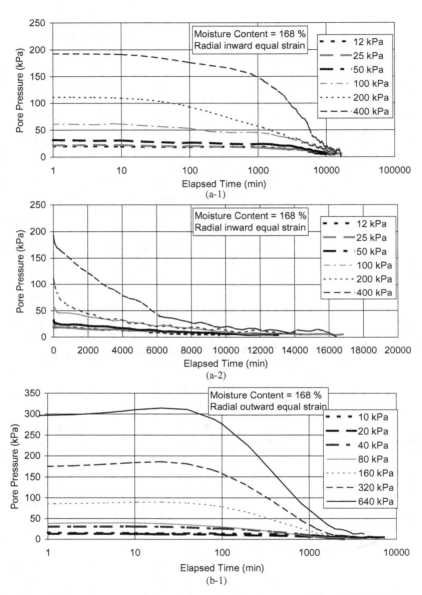

Fig. 7.24. Pore pressure versus time from various loading steps ($W = 168\%$). (a) Inward drainage, (b) outward drainage, (c) inward and outward drainage. (1 and 2 refer to semi-log scale and arithmetic scale, respectively.)

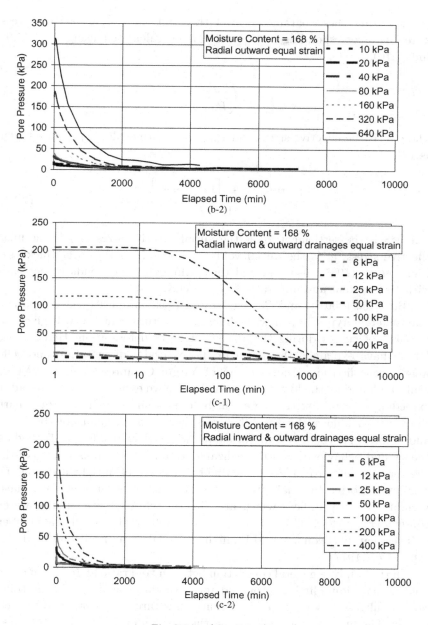

Fig. 7.24. (Continued).

By using these parameters and the void ratio e at various stress states, the corresponding settlement can be calculated using Eqs. (7.9) and (7.10).

$$e = \alpha \left(\frac{\sigma'}{P_{\text{atm}}}\right)^\beta + \varepsilon \tag{7.9}$$

where σ' is the effective stress and P_{atm} is the atmospheric pressure taken as 101.3 kPa.

$$S = \frac{\alpha \left(\frac{\sigma'}{P_{\text{atm}}}\right)^\beta \left[1 - \left(\frac{\sigma'_f}{\sigma'_i}\right)^\beta\right]}{1 + e_i} \times H \tag{7.10}$$

where S is the settlement; σ'_i is the initial effective stress; σ'_f is the final effective stress; e_i is the initial void ratio, and H is the layer thickness. Hight et al. (1987) also reported that stiffness of clay tends to decrease with clay content and mineral activity of clay.

Bo et al. (2001b) and Carrier III and Beckman (1984) have reported that different types of soil with different moisture contents will give rise to a fan of curves during the viscous stage of deformation. Hight et al. (1987) have explained that even in the natural sedimentation and self-weight consolidation process a fan of Virgin Compression Lines (VCL) could be found at the low stress level. They have explained that only after exceeding a certain stress there will be only a single VCL in which part the behavior follows the Terzaghi stress dependent normalizable behavior and before that there is a fan of curves which does not follow Terzaghi's theory and behavior is non-normalizable. Countless compression curves in the very low effective stress range due to variation of initial water content was also reported by Imai (1981). At a given stress level, a clay can have variable liquidity indices which are directly related to the initial water content (Fig. 7.25). This observation was well established by a number of investigators (Leonards and Ramiah, 1960; Olson and Mitronovas, 1962; Woo et al., 1977).

In this study, a general correlation between void ratio at 10 kPa and the void ratio at liquid limit is proposed and is validated against different soils. For the ultra-soft soil, the e-log σ'_v curve has three segments as discussed earlier. The first segment falls within the first log cycle covering the stress range between 1 and 10 kPa. This is where viscous deformation took place with negligible pore pressure dissipation. Following the notations adopted

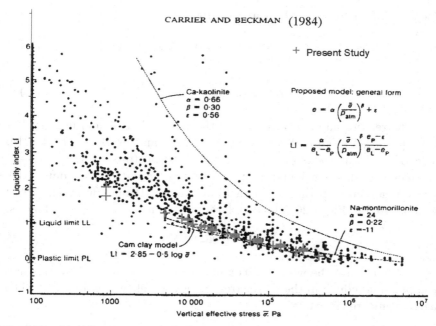

Fig. 7.25. Liquidity versus vertical effective stress for various types of clay (after Carrier and Beckman, 1984) together with tested clay from the present study.

by Burland (1990), the compression index for the first segment can be expressed as described in Eq. (7.4):

The second segment covers the stress range between 10 and 100 kPa. This is the transition zone where the soil behavior is between a viscous soil and an intrinsic soil.

$$C_{c2}^* = e_{10}^* - e_{100}^* \qquad (7.11)$$

where C_{c2}^* is the compression index in second log cycle and e_{100}^* is the void ratio at 100 kPa. The third segment covers the stress range between 100 and 1000 kPa, where the intrinsic soil behavior prevailed.

$$C_{c3}^* = e_{100}^* - e_{1000}^* \qquad (7.12)$$

where C_{c3}^* is Burland's (1990) intrinsic compression index and e_{1000}^* is the void ratio at 1000 kPa. The values of C_{c3}^* and e_{100}^* can be computed using the correlations proposed by Burland (1990).

$$e^*_{100} = 0.109 + 0.679e_L - 0.089e_L^2 + 0.016e_L^3 \quad (7.13)$$

$$C^*_{c3} = 0.256e_L - 0.04 \quad (7.14)$$

The relationship between e_{10} and the void ratio at liquid limit e_L has been proposed in an earlier section (Eq. (7.5)) and by Bo et al. (2001b). The α value can be obtained from consolidation tests on the ultra-soft samples with different liquid limits. Results of the tests carried out in this study together with the test data from Carrier III and Beckman (1984) were plotted in Fig. 7.26. The correlation between e^*_{10} and e_L is as follows:

$$e^*_{10} = 0.9641e_L \quad (7.15)$$

A comparison between the computed and measured void ratios and compression indices in the three stress ranges are shown in Table 7.8. The agreements are favorable.

7.2.4. *Validation of Basic Equation for Magnitude of Settlement*

The settlement of an ultra-soft soil can be determined using the following equations, which modified from Terzaghi's one-dimensional consolidation

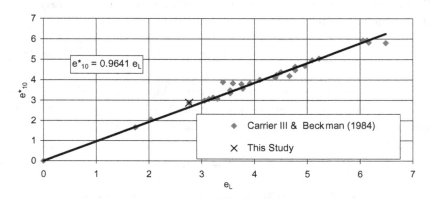

Fig. 7.26. Relationship between e^*_{10} versus e_L.

equation. For an applied pressure greater than 100 kPa, the equation is

$$S = \frac{H_0}{1+e_i}\left(C^*_{c1} + C^*_{c2} + \left(C^*_{c3} \times \log \frac{\sigma'_f}{100}\right)\right) \quad (7.16)$$

where H_0 is initial thickness of the soil.

For an applied pressure greater than 10 kPa but less than or equal to 100 kPa, the equation is

$$S = \frac{H_0}{1+e_i}\left(C^*_{c1} + \left(C^*_{c2} \log \frac{\sigma'_f}{10}\right)\right) \quad (7.17)$$

It should be noted that even though C^*_{c1} is a function of e^*_{10} in Eq. (7.4), a slight overestimation or underestimation of e^*_{10} will not affect the magnitude of the computed ultimate settlement for σ'_f greater than 20 kPa. This is because the value of C^*_{c2} is also a function of e^*_{10}. The error in C^*_{c2} will compensate the error in C^*_{c1}.

Alternatively, the ultimate settlement can be more accurately determined using the transition pressure σ'_t. For an applied pressure greater than 100 kPa, the following equation can be used. The initial stress is taken as 1 kPa.

$$S = \frac{C^*_{c1}}{1+e_i}H_0\log\sigma'_t + \frac{C^*_{c2}}{1+e_i}H_0\log\frac{100}{\sigma'_t} + \frac{C^*_{c3}}{1+e_i}H_0\log\frac{\sigma'_f}{100} \quad (7.18)$$

For an applied pressure greater than σ'_t but less than 100 kPa, the equation is

$$S = \frac{C^*_{c1}}{1+e_i}H_0\log\sigma'_t + \frac{C^*_{c2}}{1+e_i}H_0\log\frac{\sigma'_f}{\sigma'_t} \quad (7.19)$$

The transition pressure can be determined from the slurry consolidation tests or estimated from empirical correlations. According to Bo et al. (2001b), the transition void ratio e_t is approximately the same as the void ratio at liquid limit. With e_t known, the transition pressure can be readily determined from the compression test.

Table 7.9 shows the comparison of the predicted ultimate settlements based on compression parameters interpreted from void ratio at liquid limit with the measured final settlements. For completeness, the computed settlements using Carrier III and Beckman's (1984) approach are also included in Table 7.9. Figures 7.27 and 7.28 show the comparison of the measured and predicted void ratio using Eqs. (7.16) and (7.17). The method

Table 7.9. Comparison of predicted and measured settlements.

Moisture content (%)	Initial void ratio	Applied pressure (kPa)	Predicted settlement (mm) (present study)	Measured settlement (mm) (radial outward)	Measured settlement (mm) (radial inward)	Measured settlement (mm) (radial inward and outward)	Predicted Settlement (mm) (Carrier and Beckman 1984)
150	4.046	12	10.713	11.200	12.262		16.728
		25	13.756	13.272	16.572		20.634
		50	16.629	14.919	18.113		23.965
		100	19.503	17.111	20.254		26.982
		200	21.093	19.370	22.089		29.715
		400	22.684	20.985	23.330		32.180
150	4.046	14	11.350			11.990	17.580
		18	12.390			13.100	18.940
		25	13.760			14.710	20.630
		50	16.630			16.370	23.960
		100	19.540			18.580	27.020
		200	21.090			20.550	29.710
		400	22.680			22.250	32.190
168	4.513	6	10.652			9.852	11.572
		12	13.161		12.129	12.668	15.311
		25	15.819		16.218	15.320	15.819
		50	18.328		19.152	18.180	21.935
		100	20.838		21.153	20.281	24.697
		200	22.294		22.685	21.948	27.197
		400	23.750		24.037	23.751	29.462
168	4.513	10	12.502	11.893			14.363
		20	15.011	14.863			17.838
		40	17.521	18.006			20.986
		80	20.030	20.349			23.837
		160	21.826	22.296			26.418
		320	23.281	24.282			28.756
		640	24.737	25.977			30.874

presented in this paper yield better agreement with the measurements. Carrier III and Beckman's (1984) method overestimated the settlement and predicted a decrease in the settlement with an increase in the initial moisture content which is contrary to the test results. Figure 7.29 shows a comparison between the measured and calculated settlement using the proposed equations and parameters.

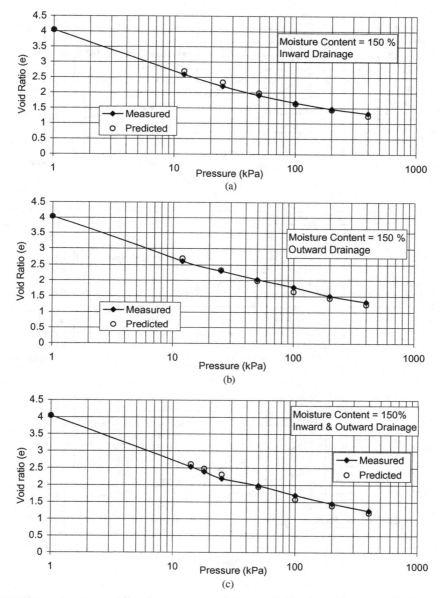

Fig. 7.27. Comparison of predicted and measured void ratio with pressure ($W = 150\%$). (a) Inward drainage, (b) outward drainage, (c) inward and outward drainage.

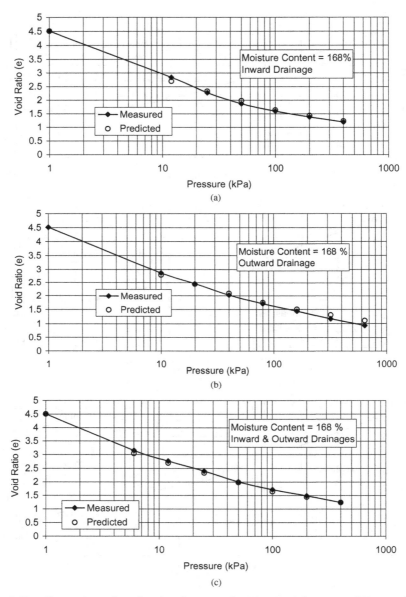

Fig. 7.28. Comparison of predicted and measured void ratio with pressure ($W = 168\%$). (a) Inward drainage, (b) outward drainage, (c) inward and outward drainage.

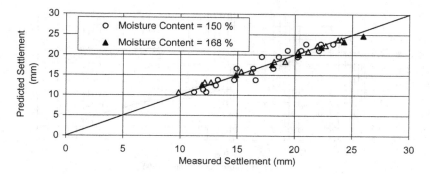

Fig. 7.29. Comparison of predicted and measured settlements.

7.2.5. Coefficient of Consolidation for Large Strain

The coefficient of hydraulic conductivity at various stages of compression can be worked out by applying the following equation for the case with inner and outer drain since the quantity of water drained out was measured:

$$k = 0.26 \frac{\Delta Q}{H \times \Delta t \times \Delta p} \log_e \frac{D}{d} \times 10^{-4} \quad \text{(m/s)} \tag{7.20}$$

where ΔQ is the volume change or the volume of water drained out; Δt is the differential time step; Δp is the differential pressure; H is the average sample thickness; D is the sample diameter, and d is the diameter of drainage filter. In this equation, the compressibility of soil was taken into consideration by updating the sample thickness in every time step.

Alternatively, k can be obtained from empirical correlations such as those with void ratio. For soils tested in this study, the correlation between the void ratio and hydraulic conductivity is shown in Figure 7.30. It should be noted that there is more or less linear relationship between the tested soil hydraulic conductivity and the void ratio of less than 3.

$$k = \exp\left(\frac{e^{-8.291}}{0.3155}\right) \tag{7.21}$$

Since the hydraulic conductivity, the stress changes, and the void ratio changes are known, the coefficient of consolidation for large strain C_F can be computed using the equation proposed by Gibson et al. (1967).

$$C_F = \frac{k}{\gamma_w(1+e)} \frac{\delta\sigma'}{\delta e} \tag{7.22}$$

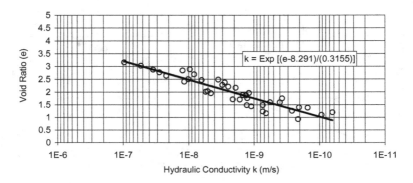

Fig. 7.30. Relationship between void ratio and hydraulic conductivity.

Although the coefficient of hydraulic conductivity changes drastically in the slurry stage, the C_F values are more or less constant as reported by Been and Sills (1981). Therefore, the C_F value can be taken as a constant for simplicity at least for each load step. Been and Sills (1981) proposed the following relationship between C_F and a dimensionless time factor:

$$T = \frac{C_F t}{D_e^2} \quad (7.23)$$

where T is the time factor; t is the time, and D_e is the drainage path.

Various dimensionless time factor curves can be obtained for various drainage conditions. Similar curves for deposition of soil including self-weight consolidation for vertical drainage at the top and bottom have been described by Lee and Sills (1981). The time factor curves for inward radial drainage determined from the settlement data for the different loads are shown in Figs. 7.31 and 7.32. It can be seen that all data clustered into one curve independent of the load magnitude. Figure 7.33 shows that the time factor curve is also independent of the moisture content for the same drainage condition.

However, outward drainage and inward drainage conditions yielded different curves in this particular tests as shown in Fig. 7.34. Inward plus outward drainage time factor curve is similar to outward drainage time factor curve due to insignificant flow contribution from inward drainage test. It can also be seen in Figs. 7.20(a) and 7.20(b) that settlement time curves of radial inward and both outward plus inward are very similar.

Lacerda et al. (1995) study of consolidation test on natural soil with the hydraulic cell also showed that outward drainage was faster than inward

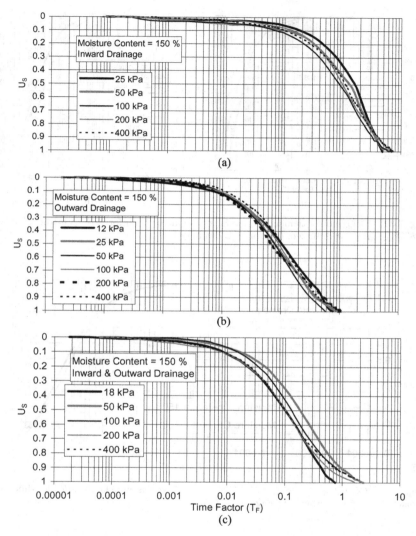

Fig. 7.31. Degree of settlement (U_s) versus time factor (T_F) for various loadings ($W = 150\%$). (a) Inward drainage, (b) outward drainage, (c) inward and outward drainage.

drainage. It should be noted that inward drainage time factor is largely dependent upon the diameter ratio of the central drain. Equations (7.24)–(7.26) can be used to determine the degree of settlement U_s with a known time factor.

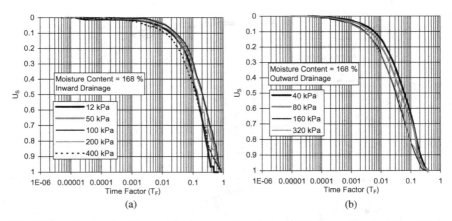

Fig. 7.32. Degree of settlement (U_s) versus time factor (T_F) for various loadings ($W = 168\%$). (a) Inward drainage, (b) outward drainage.

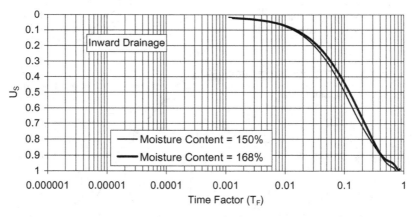

Fig. 7.33. Comparison of degree of settlement (U_s) versus time factor (T_F) for different moisture contents (inward drainage).

For inward drainage:

$$U_s = 0.2304 \text{ Ln}(T_F) + 1.0402 \tag{7.24}$$

For outward drainage:

$$U_s = 0.2186 \text{ Ln}(T_F) + 1.2811 \tag{7.25}$$

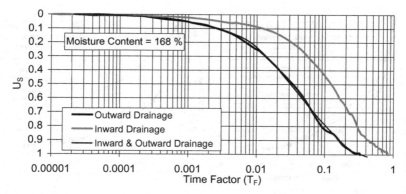

Fig. 7.34. Comparison of degree of consolidation (U_s) versus time factor (T_F) for various drainage conditions ($W = 168\%$).

For inward and outward drainage:

$$U_s = 0.2196 \, \text{Ln}(T_F) + 1.2784 \tag{7.26}$$

With the known time factor, the time rate of settlement can be readily determined. The comparison of the time rate of settlement calculated using the proposed time factor curves and the measured data for the various load steps are shown in Figs. 7.35–7.39. The predicted curves are in reasonable agreement with the measured values.

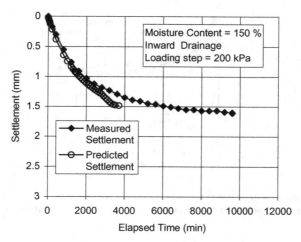

Fig. 7.35. Comparison of predicted and measured settlements with time, $W = 150\%$ (inward drainage).

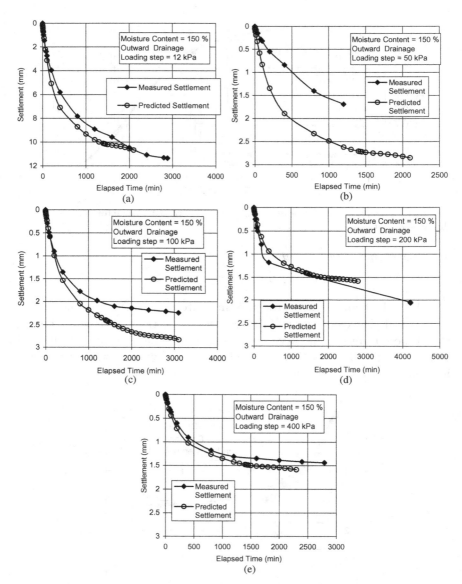

Fig. 7.36. Comparison of predicted and measured settlements with time ($W = 150\%$) outward drainage. Loading steps = (a) 12 kPa, (b) 50 kPa, (c) 100 kPa, (d) 200 kPa, (e) 400 kPa.

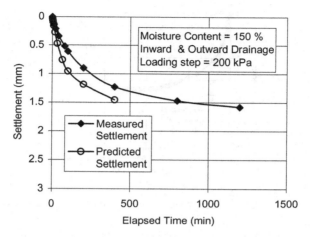

Fig. 7.37. Comparison of predicted and measured settlements with time, $W = 150\%$ (inward and outward drainage, loading step = 200 kPa).

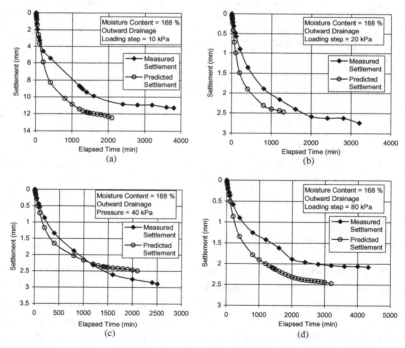

Fig. 7.38. Comparison of predicted and measured settlements with time ($W = 168\%$, outward drainage). Loading steps = (a) 10 kPa, (b) 20 kPa, (c) 40 kPa, (d) 80 kPa, (e) 160 kPa.

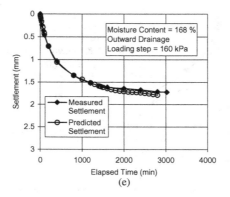

Fig. 7.38. (*Continued*).

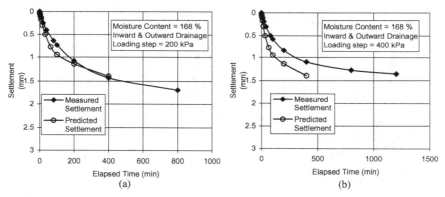

Fig. 7.39. Comparison of predicted and measured settlements with time ($W = 168\%$, inward and outward drainage). Loading steps = (a) 200 kPa, (b) 400 kPa.

7.3. Summary

(i) *One-Step High Pressure Loading Tests*

Results of tests with shorter drainage path of 40 mm also confirmed the existence of a period where large deformation takes place with little pore pressure dissipation when an external load was applied to an ultra-soft soil in the slurry stage. At the end of this period, the corresponding effective stress was about 10 kPa. Compression index in the viscous stage (C_{c1}^*) was very high. The higher the initial moisture content, the greater was the C_{c1}^*. Compression index in the soil stage (C_{c2}^*) measured between e_{10} and e_{100}

is slightly higher than Burland's (1990) intrinsic compression index which was measured between e^*_{100} and e^*_{1000}.

The transition points can be determined based on pore pressure, settlement rate, hydraulic conductivity change rate, and e-log σ'_v curve using a graphical approach. The void ratio at the transition point was approximately the same as that at the liquid limit. However, the transition void ratios decrease with increasing applied loads and increasing initial moisture contents.

(ii) *Step Loading Tests*

By carrying out the step loading compression tests on an ultra-soft soil in the hydraulic consolidation cell, the variations of void ratio with effective stress can be obtained. From the e-log σ'_v curves, the compression indices (C^*_{c1}, C^*_{c2}, and C^*_{c3}) for the three stress ranges (1–10 kPa, 10–100 kPa, and 100–1000 kPa) can be determined. Alternatively, these indices can be computed using the proposed correlations between the void ratio at 10 kPa (e^*_{10}) and the void ratio at the liquid limit as well as Burland's (1990) equations to obtain e^*_{100} and the intrinsic compression index (C^*_{c3}). The predicted settlements using these correlations agreed well with the measured values.

By measuring the volume change and pore pressure in the hydraulic cell, the hydraulic conductivity, void ratio changes, and effective stress changes as well as the coefficient of consolidation for large strain C_F can be readily determined. Alternatively, the C_F values can be estimated using empirical correlations between hydraulic conductivity and void ratio and the correlations between e^*_{10}, e^*_{100} and compression indices with the void ratio at the liquid limit.

The dimensionless time factors for the three different drainage conditions have been established using the test data. The time rate of settlement can be determined using the proposed U_s-versus-T_F curves. It was found that the predicted time rate of settlement agreed reasonably with the measured data for most of the stress ranges.

Chapter 8

CONTINUOUS LOADING TESTS ON ULTRA-SOFT SOIL

8.1. Constant Rate of Loading Test on Ultra-soft Soil

Consolidation tests on ultra-soft soil took a longer time than the natural soil due to viscous effect in the early stage. More than 5000 mins may be required to reach the end of primary consolidation for each load increment using the hydraulic consolidation cell with radial drainage condition. The constant rate of loading (CRL) tests could be carried out to investigate the possibility of characterizing the compressibility of the ultra-soft soil in the slurry stage including the deformation and pore pressure behavior during the CRL. Different test methods for applying and controlling the axial load on natural clay have been described by Aboshi *et al.* (1970), Irwin (1975), and Burghignoli (1979). In the present study, the Rowe cell was used and details of the apparatus and method of testing are explained in the following section.

8.1.1. *Apparatus Used*

The apparatus used was the Rowe cell described in the previous chapter. The only difference is that a small diameter Rowe cell (75 mm) was used in the CRL test and the samples were 27–30 mm thick (Fig. 8.1). The preparation of samples has been described in an earlier chapter. Drainage was allowed only at the top and the pore pressure was measured at the center of the base. The CRL tests were carried out on samples with two different moisture contents subjected to various loading rates. The details are summarized in Table 8.1. The loading rates varied from 1 kPa/150 s to 1 kPa/1500 s (i.e., 2.4 kPa/h to 24 kPa/h). These rates are far slower than those used by Aboshi *et al.* (1970) who used a rate of 300 kPa/h and are

Fig. 8.1. Photographic features of CRL and CRS tests.

Table 8.1. Details of tests.

Test code	Moisture content (%)	Initial void ratio	Rate of loading
150	147.65	4.038	1 kPa/300 s
	149.22	4.011	1 kPa/500 s
	150.77	4.040	1 kPa/800 s
	150.02	4.046	1 kPa/1500 s
170	169.31	4.649	1 kPa/150 s
	168.01	4.530	1 kPa/300 s
	167.98	4.528	1 kPa/500 s
	173.22	4.676	1 kPa/800 s

closer to that used by Irwin (1975) at 6 kPa/h using 150 mm diameter Rowe cell. For a test carried out up to 400 kPa, it can take from 17 h to 7 days for a rate of 1 kPa/150 s and 1 kPa/1500 s, respectively. The time taken for the tests using the slowest rate is still faster than the time required for the conventional consolidation test on the same thickness of a natural soil sample. The applied pressure was provided by a GDS controller which allows a steady increase in stress on the specimen. The pore pressure was measured at the base of the sample. A back-pressure of 100 kPa was applied in all tests. The vertical settlement was measured by a LVDT installed on top of the piston rod.

8.1.2. Discussion of the Results

8.1.2.1. Settlement and Pore Pressure Behavior

It can be seen in Fig. 8.2 that for a soil with the same moisture content both the magnitude as well as the rate of settlement vary with different rates of loading. The faster the loading rate the greater is the magnitude and rate of settlement due probably to a faster effective stress gain during the test. The ultimate settlement of the same moisture content soil is also affected by the loading rate. The faster the loading rate the greater is the magnitude of ultimate settlement (Figs. 8.2(a) and 8.2(b)). The effect of loading on the magnitude of the final settlement has been reported by Matsuda and Aboshi (1993) and Matsuda and Nagatani (1995). However, for a similar loading rate the 170% moisture content soil settles faster and greater than the 150% moisture content soil (Figs. 8.3(a)–8.3(c)).

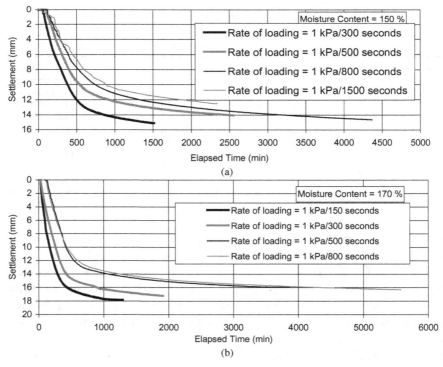

Fig. 8.2. Comparison of settlement versus time from various rate of loadings. (a) $W = 150\%$, (b) $W = 170\%$.

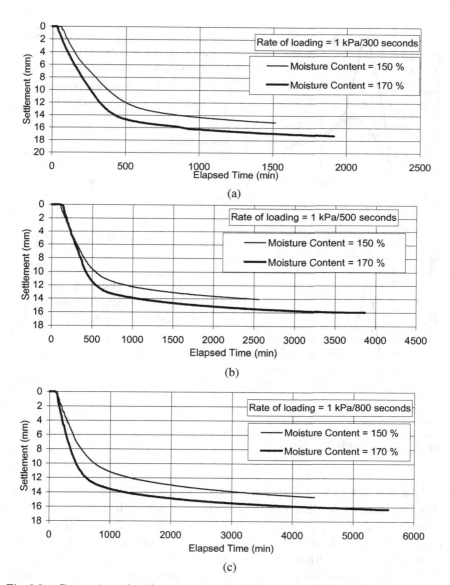

Fig. 8.3. Comparison of settlement versus time between two different moisture content soils under same rate of loading. Rate of loading = (a) 1 kPa/150 s, (b) 1 kPa/500 s, (c) 1 kPa/800 s.

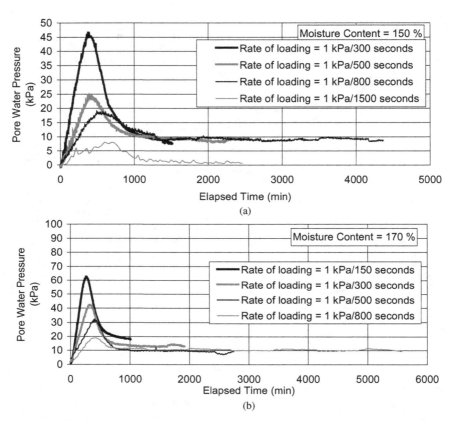

Fig. 8.4. Comparison of pore pressure versus time from various rate of loadings. (a) $W = 150\%$, (b) $W = 170\%$.

Figures 8.4(a) and 8.4(b) show the variation of pore pressure with time for various loading rates for tests on soils with 150% and 170% moisture content. The pore pressure increases together with the cell pressure to a peak at 380–720 min for the sample with 150% moisture content and 270–420 min for the sample with 170% moisture content. The pore pressure falls to about 10–15 kPa and remained constant until the end of the tests (Figs. 8.4(a) and 8.4(b)). The highest peak pore pressure recorded was between 47–62 kPa in the fastest rate of loading, whereas the lowest peak pore pressure was recorded as 8–17 kPa at the slowest rate of loading for samples with 150% and 170% moisture content, respectively. In general, approximately the same magnitude of peak pore pressure was reached although in the higher moisture content test, slightly lower peak pore pressures occurring in slightly shorter time was observed (Figs. 8.5(a)–8.5(c)). The largest magnitude and fastest rate of settlement were recorded during the period of increasing pore pressure. As the pore pressure dissipated, smaller and slower rates of settlement occurred.

The time taken to reach the peak pore pressure varies with the rate of loading. The pore pressures peaked faster with faster rate of loading. The value of pore pressure, cell pressure, settlement, and void ratio at peak are shown in Table 8.2. It can be seen in the table that the peak pore pressure was higher for the faster rate of loading. At the time when pore pressure reached the peak values, pore pressure ratios (u_b/σ) at the base which is ratio between pore pressure measured at the base and applied pressure were ranging between 0.28–0.60 for 150% moisture content tests and 0.53–0.60 for 170% moisture content tests. Generally pore pressure ratio (u_b/σ) is increasing with loading rate. However, except for $1\,\text{kPa}/1500\,\text{s}$ tests, the (u_b/σ) ratio falls within a narrow range of 0.51–0.60. At the time when the pore pressure reached the peak, a narrow range of settlement was registered which varied between 9.7–13.56 mm for 170% moisture content and 7.7–10 mm for 150% moisture content. These settlements correspond to a strain of 28–45%, which increases with the rate of loading. The void ratios at peak pore pressure are found in the range between 2.1 and 2.68.

8.1.2.2. Effective Stress Gain at the Base and Average Effective Stress

It has been reported by various researchers that the average effective stress gain in the soil is greater than the gain at the base due to the variation of pore pressure along the soil profile. Crawford (1965) derived the average effective stress gain assuming average pore pressure as one half of pressure at the base whereas Smith and Wahls (1969) used factors varying between 0.667 and 0.75 depending on the dimensionless ratio of b/r which in turn is dependent upon the void ratio profile that varied between 0 and 2 in which b is a constant that depends on the variation in void ratio with depth and time and r is the rate of void ratio change. Aboshi et al. (1970) calculated the average effective stress gain from dimensionless time factor curves which were modified from Schiffman (1958) equations. Schiffman (1958) described the excess pore pressure from continuous loading as a function of the time

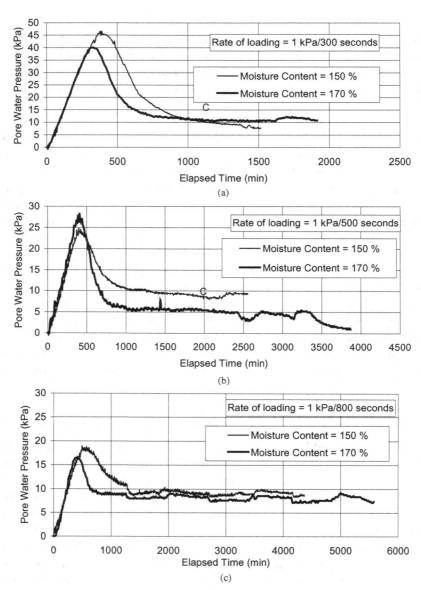

Fig. 8.5. Comparison of pore pressure versus time between two different moisture content soils under same rate of loading. Rate of loading = (a) 1 kPa/300 s, (b) 1 kPa/500 s, (c) 1 kPa/800 s.

Table 8.2. Peak pore pressure and void ratio.

Moisture content (%)	Loading rate	Pore pressure at peak (kN/m^2)	Void ratio at peak	Elapsed time at peak (min)	u_b/σ at peak	Settlement at peak (mm)	Strain at peak (%)	Void ratio at liquid limit
150	1 kPa/300 s	46.60	2.64	390	0.597	10.24	28	2.76
150	1 kPa/500 s	24.80	2.32	400	0.517	8.09	34	
150	1 kPa/800 s	19.00	2.51	495	0.512	7.69	30	
150	1 kPa/1500 s	8.00	2.63	720	0.278	7.67	28	
170	1 kPa/150 s	61.80	2.10	270	0.594	13.56	45	
170	1 kPa/300 s	40.00	2.32	330	0.606	12.20	40	
170	1 kPa/500 s	28.20	2.57	410	0.573	9.77	35	
170	1 kPa/800 s	16.60	2.68	420	0.527	9.67	35	

factor T and the location of the element as follows:

$$u = \frac{16u_0}{T_0\pi^3} \sum_{n=1,3,6...}^{n=\alpha} \frac{1}{n^3} \sin\frac{n\pi z}{2H} \left\{1 - \exp\left[-\left(\frac{n^2\pi^2}{4}\right)\right]T\right\} \quad (8.1)$$

where u = imposed excess pore pressure; T_0 = time factor arbitrarily determined on the linear loading curve; and u_0 = excess pore pressure which may be exerted by the instantaneous loading of total pressure σ at T_0.

The value u_0/T_0 in Eq. (8.1) is the rate of loading and is an arbitrarily chosen. Aboshi *et al.* (1970) transformed Schiffman's expression to

$$\frac{u}{\sigma} = \frac{16}{\pi^3 T} \sum_{n=1,3,5...}^{n \to \alpha} \frac{1}{n^3} \sin\frac{n\pi z}{2H} \left\{1 - \exp\left[-\left(\frac{n^2\pi^2}{4}\right)T\right]\right\} \quad (8.2)$$

in which σ is the external load at T. This expression is the function of only two variables: factor T and element location z/H. Therefore, the excess pore pressure ratio at the impermeable base (u_b/σ) is defined as

$$\frac{u_b}{\sigma} = \frac{16}{\pi^3 T} \sum_{m=0}^{\alpha} (-1)^m \left\{\frac{1}{(2m+1)^3}\right\} \left\{1 - \exp\left[\frac{-(2m+1)^2\pi^2}{4}T\right]\right\} \quad (8.3)$$

where u_b is the base pore pressure and average pore pressure ratio could be expressed as follows:

$$\frac{\bar{u}}{\sigma} = \frac{32}{\pi^4 T} \sum_{m=0}^{\alpha} \frac{1}{(2m+1)^4} \left\{1 - \exp\left[\frac{-(2m+1)^2\pi^2}{4}T\right]\right\} \quad (8.4)$$

Hence the average degree of consolidation in terms of pore pressure can be obtained below

$$U = 1 - \frac{\bar{u}}{\sigma} \quad (8.5)$$

The average effective stress σ' at a particular time is

$$\sigma' = U \times \sigma \quad (8.6)$$

However, if the ratio of average excess pore pressure to excess pore pressure at the base is calculated from (8.3) and (8.4), it can be seen that for the time factor ranges between 0.1 and 10, ratios are found to be between 0.667 and 0.77. The ratio is highest at the lower time factor and it is constant at 0.667 when the time factor is greater than unity (Fig. 8.6).

Therefore, the average effective stress gain calculated can be simplified by applying 0.667 of base pore pressure for time factors greater than unity.

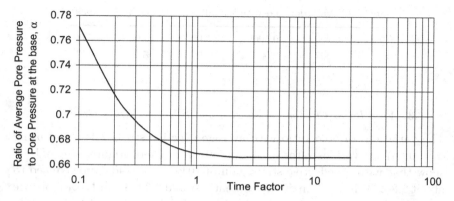

Fig. 8.6. Ratio of average pore pressure to pore pressure at the base versus time factor.

Smith and Wahls (1969) reported that the average effective stress gain from continuous loading for tests carried out without back-pressure can be obtained from the following equation:

$$\sigma'_1 = \sigma_1 - \alpha u_b \qquad (8.7)$$

where σ'_1 is the average effective stress gain; σ_1 is the applied stress; u_b is the pore pressure at the base; and α is the ratio of average pore pressure to pore pressure at the base.

The ratio of average pore pressure can be obtained from (8.8).

$$\alpha = \frac{\bar{u}}{u_b} = \frac{\frac{1}{3} - \frac{b}{r}\left(\frac{1}{24}\right)}{\frac{1}{2} - \frac{b}{r}\left(\frac{1}{12}\right)} \qquad (8.8)$$

where $\bar{u}$ is the average pore pressure; and b/r is the dimensionless ratio which varies between 0 and 2.

Smith and Wahls (1969) reported that when $b/r = 0$, it implied that the void ratio is uniform with depth and the α value is equal to 0.667 or 2/3. When $b/r = 2$, there is no change in void ratio at the base in which the α value is 0.75. The variation of α value with b/r is shown in Table 8.3. It can be seen that at the early stage of compression with lower time factor values, the variation of void ratio may be higher. As such the α values could be higher. However, majority of the values are found to be 0.667.

Janbu et al. (1981) and Leroueil et al. (1985) reported the same value of 0.667 for the constant rate of strain (CRS) test on natural soil for u_b/σ lower than 0.4. The α values become higher with increasing u_b/σ values. Leroueil et al. (1985) reported α values as high as 0.75 for $u_b/\sigma = 0.92$.

Table 8.3. Variation in α with b/r (after Smith and Wahls, 1969).

b/r	0.000	0.500	1.000	1.500	2.000
α	0.667	0.682	0.700	0.722	0.750

8.1.2.3. Void Ratio Change versus Average Effective Stress Gain

The variations of void ratio with average effective stress calculated using $\alpha = 0.667$ for various rates of loading are compared. Figures 8.7 and 8.8 show that with an effective stress gain of 10 kPa the void ratio reduced to about 2.8–2.97 in 150% moisture content test and 2.93–3.2 in 170% moisture content test. The rate of void ratio change for the same moisture content soil is a function of the rate of loading. The slower the loading rate the greater is the change in void ratio in the first log cycle up to 10 kPa effective stress. Beyond 20 kPa, the e-log σ' curves converged although there were slight variations in the final void ratio. At 20 kPa, the void ratio was found to be about 2.4–2.5 for both moisture content soils. The compressibility parameters are compared in Table 8.4 and discussed in the subsequent section.

8.1.2.4. Compression Indices

For the e-log σ' of slurry compression, the curve has a slight curvature with decreasing magnitude of compression as the stress increases. Since a non-linear relationship is difficult to use in the analysis, the curve was simplified into three linear sections with one each log cycles (SI unit). There are three compression indices in the e-log σ' curve of the slurry compression. The compression index is highest in the 1st log cycle and then decreases in subsequent log cycles. Similar behavior was found in the CRL compression curves. C_{c3}^* is very similar to Burland's (1990) intrinsic compression index and C_{c2}^* is slightly greater than C_{c3}^*. C_{c3}^* is more or less the same for all rates of loading as well as the different moisture contents. However, C_{c2}^* varied slightly between different rates of loading as well as different moisture contents due to slight variations in e_{10}^* values. The slower rate of loading gives slightly lower C_{c2}^*. The C_{c1}^* values decreases with reducing loading rates. The measured compression indices C_{c1}^*, C_{c2}^* and C_{c3}^* at various stages and the values of e_{10}^* and e_{100}^* are compared with those calculated using Eqs. (7.4), (7.13), and (7.15). It was found that the calculated values were very close to the measured values of C_{c3}^* and

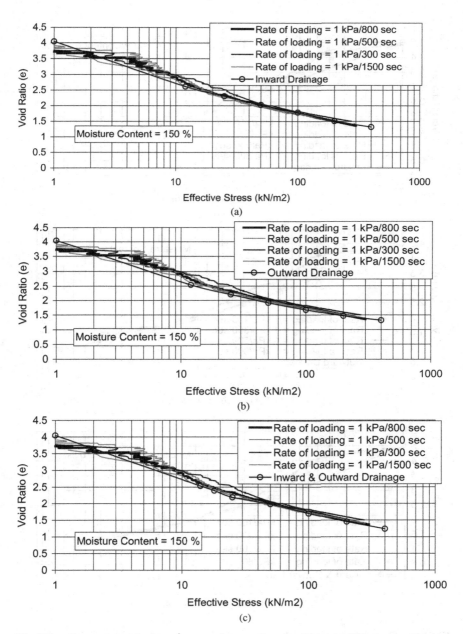

Fig. 8.7. Comparison of e-log σ' curves from various loading rate CRL tests with step loading hydraulic cell tests ($W = 150\%$). (a) Inward drainage, (b) outward drainage, (c) inward and outward drainage.

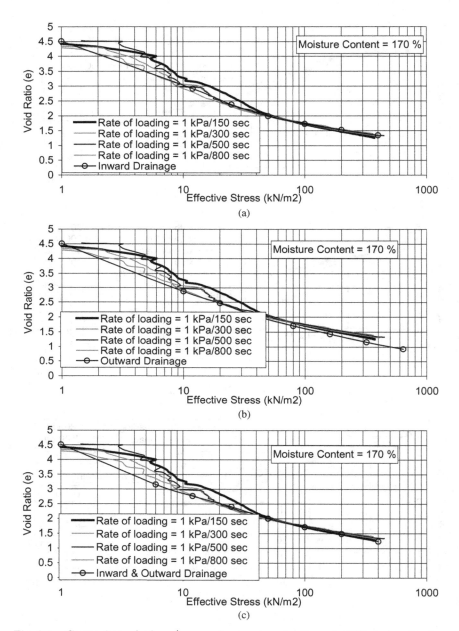

Fig. 8.8. Comparison of e-log σ' curves from various loading rate CRL tests with step loading hydraulic cell tests ($W = 170\%$). (a) Inward drainage, (b) outward drainage, (c) inward and outward drainage.

Table 8.4. Comparison of measured and calculated compression parameters ($\alpha = 0.667$).

Moisture content (%)	Initial void ratio	Loading rate	Measured						Calculated				
			e_{10}^*	e_{100}^*	C_{c1}^*	C_{c2}^*	C_{c3}^*		e_{10}^*	e_{100}^*	C_{c1}^*	C_{c2}^*	C_{c3}^*
150	4.038	1 kPa/300 s	2.950	1.813	1.088	1.137	0.688		2.770	1.770	1.268	0.997	0.660
150	4.011	1 kPa/500 s	2.970	1.75	1.041	1.220	0.750				1.241		
150	4.040	1 kPa/800 s	2.910	1.75	1.13	1.160	0.813				1.270		
150	4.046	1 kPa/1500 s	2.813	1.688	1.233	1.125	—				1.276		
170	4.649	1 kPa/150 s	3.200	1.725	1.449	1.475	0.775				1.879		
170	4.530	1 kPa/300 s	3.075/3.15	1.775	1.455	1.300	0.625				1.760		
170	4.528	1 kPa/500 s	2.925/3.09	1.750	1.573	1.205	0.650				1.758		
170	4.676	1 kPa/800 s	2.925	1.725	1.751	1.200	0.650				1.906		

e_{100}^*. The measured C_{c2}^* or e_{10}^* values are close to the calculated values in the slowest rate of loading in both 150% and 170% moisture content tests.

The e-log σ' curves obtained from the CRL tests are compared with those from the hydraulic cell tests carried out with various drainage conditions. It is found that both in 150% and 170% moisture content tests, there are general agreement in the second and third log cycles (Figs. 8.7 and 8.8). In the first log cycle (between 1 kPa and 10 kPa) the slope obtained from test at 1 kPa/800 s was close to the slopes of the step loading hydraulic cell tests in both 150% and 170% moisture content tests. The loading rate of 1 kPa/1500 s is found to be too slow. Therefore, the rate of 1 kPa/800 s is adopted for the ultra-soft soil test. At this loading rate, the time required is still much shorter than that of end of primary consolidation step loading test.

8.1.2.5. Variation of e-log σ' Curve with α Values

It has been speculated that the α value may vary due to variation of void ratio profile in the tested soil during the test. However, in this study the variations of e-log σ' curves from five selected α values between 0.667 and 0.75 are discussed in order to find out the suitable α values for the two moisture content soils. It can be seen in Figs. 8.9(a)–8.9(c) that the higher α values are more closely matched with the slopes from the step loading hydraulic cell test. The effect is more pronounce in low effective stress range. It agreed well with higher α values determined from the equation by Aboshi et al. (1970) in the lower range of time factor.

The measured compression parameters such as e_{10}^*, e_{100}^*, C_{c1}^*, C_{c2}^* and C_{c3}^* from conventional tests were again compared with those from selected rate of loading applying higher α values of 0.75. The parameters using $\alpha = 0.75$ are in better agreement with those from conventional tests than parameters obtained from $\alpha = 0.667$ (Table 8.5).

8.1.2.6. Hydraulic Conductivity and Large Strain Coefficient of Consolidation

The hydraulic conductivity at any time during the CRL test can be determined using the equation:

$$k = \frac{\delta v}{\delta t} \frac{\bar{H}_t \gamma_w g}{2A\delta u} \tag{8.9}$$

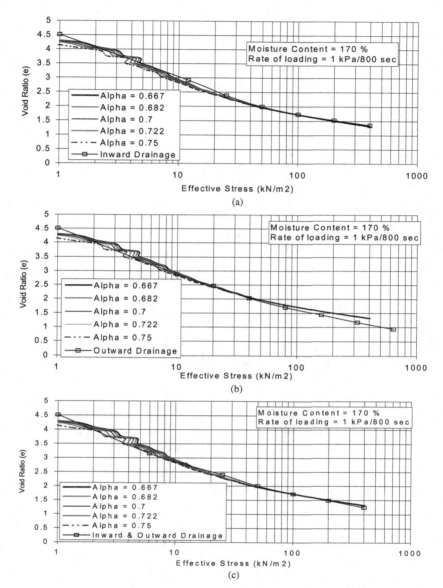

Fig. 8.9. Comparison of e-log σ' curves resulted from various α values in 1 kPa/800 s, rate of loading tests with step loading hydraulic cell test. (a) Inward drainage, (b) outward drainage, (c) inward and outward drainage.

Table 8.5. Comparison of measured and calculated compression parameters from selected rate of loading (1 kPa/800 s) applying selected α value of 0.75.

Moisture content (%)	Initial void ratio	Measured						Calculated				
		e_{10}^*	e_{100}^*	C_{c1}^*	C_{c2}^*	C_{c3}^*		e_{10}^*	e_{100}^*	C_{c1}^*	C_{c2}^*	C_{c3}^*
150	4.04	2.820	1.75	1.22	1.07	0.813		2.770	1.770	1.270	0.997	0.660
170	4.676	2.775	1.725	1.901	1.050	0.650				1.906		

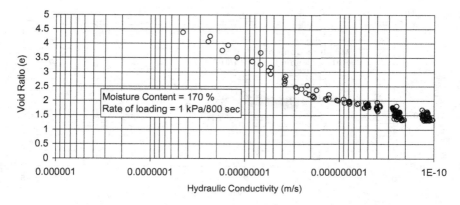

Fig. 8.10. Void ratio versus hydraulic conductivity.

where δv is the volume of water drained out from the sample between two differential time, δt is the differential time, $\bar{H}_t$ is average thickness at time t, δu is differential pore pressure between drained and undrained face, γ_w is density of water, and A is the area of sample. The compressibility of soil was taken into considered by updating sample thickness in each time step.

The large strain coefficient of consolidation can be obtained as follows:

$$C_F = \frac{k}{\gamma_w(1+e)} \frac{\delta \sigma'}{\delta e} \qquad (8.10)$$

where the average effective stress gain $\delta \sigma'$ is obtained from pore pressure measurement and void ratio change δe from monitored settlement.

As stated by Lee and Sills (1981) C_F values are found to be almost constant although other parameters are highly varied. Figure 8.10 shows the variation of hydraulic conductivity with void ratio from CRL test, whereas Fig. 8.11 shows almost constant C_F values obtained from the CRL test.

8.2. Constant Rate of Strain Test on Ultra-Soft Soil

The CRL test at 1 kPa/800 s was suggested because both step loading compression test and large-scale consolidation tests were very time consuming. CRL test can be completed within 4 days. The ultimate magnitude of settlement is however slightly affected by the rate of loading. Another alternative is to conduct the test at constant rate of strain (CRS).

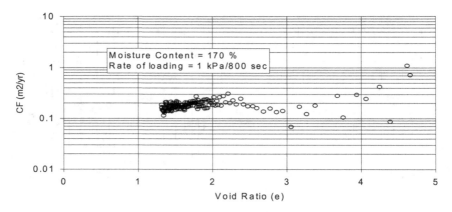

Fig. 8.11. C_F versus void ratio.

The CRS test on natural soil has been described by a number of researchers such as Smith and Wahls (1969), Anwar et al. (1971), and Gorman et al. (1978) and Sheahan and Watters (1996). Its application to very soft clayey soils such as dredged material and slurry were reported by Umehara and Zen (1980) and Carrier III and Beckman (1984). Tests on gaseous soils were reported by Wichman (1999). In this section the CRS tests on two high moisture content soils conducted at four different strain rates are discussed.

8.2.1. The Apparatus

The CRS test was carried out using a 70 mm diameter hydraulic Rowe cell. The specimen was loaded vertically under one-dimensional conditions by hydraulic pressure on a flexible cell diaphragm. The cell pressure and back pressure were separately controlled with the GDS system. The total pressure applied was measured by the pressure transducer connected to the diaphragm cell. The CRS was achieved by pumping water into the diaphragm cell at a specified rate. A displacement transducer LVDT was used to countercheck the strain rate.

8.2.2. Sample Preparation and Test Method

Soil samples were collected from a siltpond where a very soft silty clay was submerged under seawater at about 7 m depth. Samples with a thickness of

30 mm and moisture contents of 130% and 150% were prepared according to (4.2) and tested under strain rates between 0.01%/min and 0.1%/min.

The samples were prepared in the Rowe cell. A cell pressure of 51 kPa was applied with a back pressure of 50 kPa to ensure sample saturation. The following measurements were made during the tests: settlement with the LVDT system, pore pressure at the center of the base, cell pressure in the hydraulic cell, and volume change. The test was carried out with single drainage at the top.

8.2.3. Selection of Strain Rate

Various strain rates based on liquid limit, coefficient of consolidation, and excess pore pressure ratio have been proposed by several researchers. Wissa et al. (1971) used the strain rate which generated a base excess pore pressure (Δu_b) of less than 30% of applied total vertical stress (σ_v). The strain rate has been standardized in ASTM D4186-82 where values ranging between 0.0001%/min and 0.04%/min have been recommended.

Smith and Wahls (1969) proposed the strain rate for natural soil given by the following equation based on C_v and C_c;

$$R = \frac{C_v C_c}{m^2 H_0 (1+e_i)} \left[\frac{\frac{u_b}{\sigma_1}}{1 - 0.7\left(\frac{u_b}{\sigma_1}\right)} \right] \quad (8.11)$$

where m is a proportionality constant ranging between 0.6 and 0.8.

For large strain consolidation, C_F is used instead of C_v. Gibson (1958) suggested the relationship between C_F and C_v as follows:

$$C_F = \frac{C_v}{(1+e)^2} \quad (8.12)$$

Hence Eq. (8.11) becomes:

$$R = \frac{C_F (1+e)^2 C_c}{m^2 H_0 (1+e_i)} \left[\frac{\frac{u_b}{\sigma_1}}{1 - 0.7\left(\frac{u_b}{\sigma_1}\right)} \right] \quad (8.13)$$

The compression index C_{c1}^* in the first log cycle can be determined from the void ratio at 10 kPa using (7.5) and the liquid limit. Since the hydraulic conductivity at any void ratio can be estimated using (7.21), C_F can be obtained. The strain rate for low stress range which is most important for ultra-soft soil, CRS tests can be estimated using (8.13). Table 8.6 shows

Table 8.6. Suggested strain rate for high moisture content soil.

Moisture content (%)	Liquid limit (%)	Suggested strain rate applying C_c^* (%/min)	Suggested strain rate applying C_{c1}^* (%/min)
130	78	0.006	0.016
	90	0.025	0.043
	100	0.083	0.096
150	78	0.004	0.013
	90	0.013	0.035
	100	0.041	0.083
170	78	0.002	0.011
	90	0.009	0.030
	100	0.025	0.069
190	78	0.002	0.010
	90	0.006	0.027
	100	0.010	0.063

the strain rate estimated using (8.13) for soils with various liquid limits. It can be seen in the table that the greater the liquid limit the higher is the strain rate, whereas the higher the moisture content the slower is the strain rate. A comparison of the strain rate based on C_{c1}^* and C_c^* shows that the strain rates calculated from the intrinsic compression index, C_c^* is lower than that calculated from C_{c1}^*. It appears that for a high moisture content soil the strain rate required range from 0.01%/min to 0.1%/min. Therefore, the tests were carried out with strain rates ranging from 0.01%/min to 0.10%/min since the liquid limit of the selected ultra-soft soil was about 100.

8.2.4. *Discussion on the Test Results*

8.2.4.1. *Behavior of Pore Pressure and Applied Load*

As can be seen in Fig. 8.12 the strain rate applied was linear especially in strain measurement using volume change data. A slight deviation of strain measured by the LVDT in the early stage of tests was noted. This could be due to the resistance of the diaphragm in the low stress range.

The total load measured by the pressure cell increased with time. The higher the strain rate the faster is the load increment. However, the load increment is not linear as shown in Figs. 8.13(a) and 8.13(b). Unlike the CRL test, the excess pore pressure increased together with the increase in total pressure. The higher the strain rate, the greater was the pore pressure increase at the base. This is consistent with the pore pressure behavior observed in the CRS tests on natural soil reported by Leroueil et al. (1985).

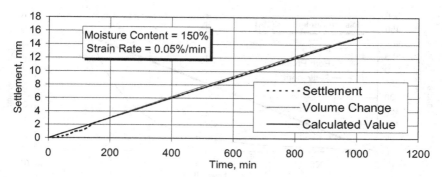

Fig. 8.12. Comparison of settlement versus time between settlement measured from LVDT, volume change, and applied strain rate.

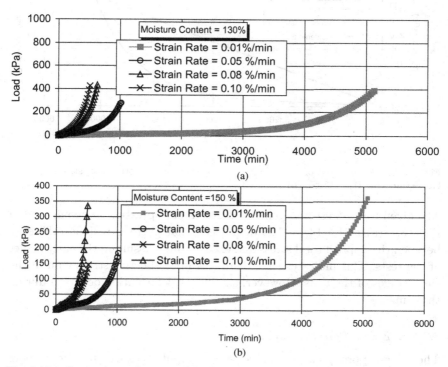

Fig. 8.13. Comparison of measured load versus time in various strain rate. (a) $W = 130\%$, (b) $W = 150\%$.

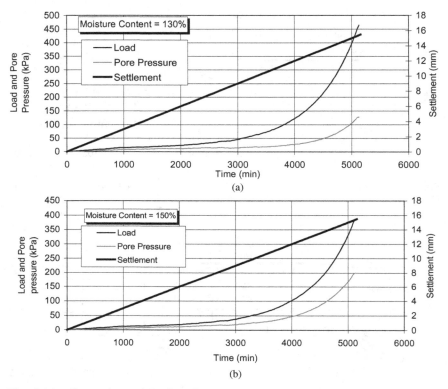

Fig. 8.14. Comparison of load and pore pressure versus time at strain rate = 0.01%/min. (a) $W = 130\%$, (b) $W = 150\%$.

The given total stress and generated excess pore pressure are compared in Figs. 8.14(a) and 8.14(b). It can be seen that there is a little difference between the applied total stress and excess pore pressure in the early stages of the tests. This indicates that there was little effective stress gain in the early stages of the tests. However, significant deformation occurred during that time.

8.2.4.2. Excess Pore Pressure Ratio

The excess pore pressure ratios were computed using the excess pore pressures measured at the base. It can be seen in Figs. 8.15(a) and 8.15(b) that the higher the strain rate the greater is the stabilized excess pore pressure ratio. The peak excess pore pressure ratio increased with increasing strain rate and generally exceeded 50–60%. Only in the slowest strain rate

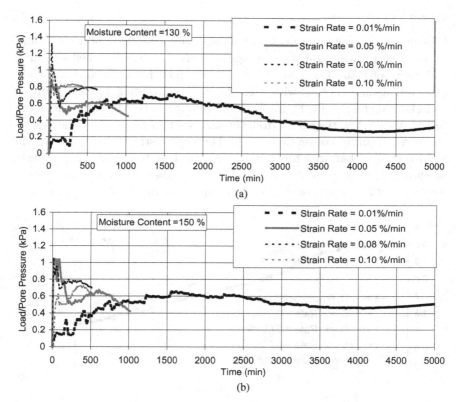

Fig. 8.15. Comparison of generated excess pore pressure ratio from various strain rate. (a) $W = 130\%$, (b) $W = 150\%$.

of 0.01%/min, the stabilized excess pore pressure ratio below 40% were noted, whereas other rates produce stabilized excess pore pressure ratios greater than 40%.

It appears that a soil with a higher moisture content requires a slower strain rate. This is consistent with the strain rate suggested by Smith and Wahls (1969). A strain rate of slower than 0.01%/min appears to be suitable for ultra-soft soil with moisture content between 130% and 150% since a reasonable steady state pore pressure can be achieved. This suggested that the suitable strain rate is slightly lower than that obtained from (8.13) which was modified from Smith and Wahls (1969). This strain rate was similar to that obtained from (8.11), which is therefore sufficient to estimate the strain rate for ultra-soft soil.

8.2.4.3. Effective Stress Gain

Effective stress gains were calculated using the measured total stress and pore pressure at the base. The average pore pressure was obtained assuming "α" value of 0.75. The e-log σ'_v curves obtained from the various strain rates are again compared with those from the conventional end of primary consolidation tests using hydraulic Rowe cell. It can be seen that among the tested strain rates, the curves from the slower strain rate test was nearest to the curves from the conventional tests. The slowest strain rate of 0.01%/min yielded results that are closely matched with these from conventional tests as shown in Figs. 8.16(a) and 8.16(b). The higher strain rate curves are well above the conventional test curve especially in low stress range when

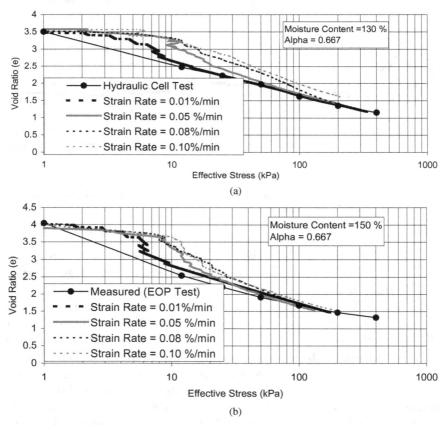

Fig. 8.16. Comparison of e-log σ'_v curves from CRS tests with various strain rate and conventional 24 h test curves. (a) $W = 130\%$, (b) $W = 150\%$.

the pore pressure ratios are high. These curves become closer at the higher stress range when the pore pressure ratio has become smaller. In the tests with high strain rate, a preconsolidation pressure was also observed which should not exist in the reconstituted ultra-soft soil. In order to obtain a reasonable e-log σ'_v curve, the slower strain rate which produces lower pore pressure ratio is therefore required. For ultra-soft soil with moisture content 130–150% having liquid limit of about 100% the CRS tests must be carried out with a strain rate of not more than 0.01%/min. It takes about 5 days to complete a test.

8.2.4.4. Compression Indices

As explained in the earlier section, the compression indices for the various stress ranges are required in order to be able to predict the magnitude of settlement at a particular stress.

It can be seen in the figures that the e-log σ' curves are affected by the rate of strain. The effect of strain rate has been explained by Leroueil et al. (1985). Leroueil (1999) attributes this to the viscous effect. The rate dependency of stress–strain behavior is also reported by Hight et al. (1987) and Jardine et al. (1991). The compression indices are also affected by the rate of strain. Only the suitable rate of strain will provide the compression index close to those obtained at the-end-of-primary-consolidation test. Smith and Wahls (1969) proposed an equation that gives a reasonable strain rate. The corresponding stabilized pore pressure is likely to be less than 30% of the applied load. Therefore, the strain rate from (8.11) can be used to obtain a reasonable e-log σ' curve.

8.2.4.5. Coefficient of Consolidation and Hydraulic Conductivity

Sheahan and Watters (1996) proposed the following equations for computing the coefficient of consolidation and hydraulic conductivity from the CRS tests for natural soil for non-transient conditions where the dimensionless time factor T is greater than 5.

$$c_v = \frac{-H^2 \log \dfrac{\sigma_{v2}}{\sigma_{v1}}}{2\Delta t \log\left(1 - \dfrac{\Delta u_b}{\sigma_v}\right)} \qquad (8.14)$$

$$c_v = \frac{-0.434\gamma H^2 \gamma_w}{2\sigma'_v \log\left(\dfrac{\sigma_v - \Delta u_b}{\sigma_v}\right)} \qquad (8.15)$$

where H is the current specimen height, γ is the strain rate and γ_w is the unit weight of water. σ_{v1} and σ_{v2} are the total stresses at two difference time Δt and σ'_v is the average effective stress obtained from:

$$\sigma'_v = (\sigma_v^3 - 2\sigma_v^2 u_b + \sigma_v u_b^2)^{1/3} \tag{8.16}$$

where u_b is the pore pressure at the base.

Smith and Wahls (1969) proposed the following equation to obtain c_v from the CRS test:

$$c_v = \frac{\gamma H^2}{a_v u_b} \left[\frac{1}{2} - \frac{b}{r}\left(\frac{1}{12}\right) \right] \tag{8.17}$$

where a_v is the coefficient of compressibility and b/r is a dimensionless ratio in which b is a constant that depends on the variation in void ratio with depth. The practical range of b/r is between 0 and 2.

Anwar et al. (1971) proposed the following equation for c_v based on non-linear theory assuming $\bar{\sigma}_v = \sigma_v$ for small u_b.

$$c_v = \frac{-0.434\gamma H^2}{2\sigma'_v m_v \log\left(1 - \dfrac{u_b}{\sigma_v}\right)} \tag{8.18}$$

In our study, the hydraulic conductivity could be calculated from time to time based on the hydraulic gradient and settlement at a particular time. Therefore, k can be expressed as

$$k = \frac{\delta h}{\delta t \times i} \tag{8.19}$$

where δh is the settlement, t is the time, and $\delta h/\delta t$ is the settlement rate. Equation. (8.19) becomes

$$k = \frac{S}{i} = \frac{S \times \gamma_w \times H}{\Delta u_m} = \frac{\delta e \times \gamma_w \times H \times H}{\Delta u_m(1 + e_0)} = \frac{\delta e \times \gamma_w \times H \times H}{\Delta u_m(1 + e_0)} \tag{8.20}$$

where Δu_m is the mean excess pore pressure.

Figures 8.17(a) and 8.17(b) shows the variation of hydraulic conductivity with void ratio from the CRS tests using (8.20). The coefficient

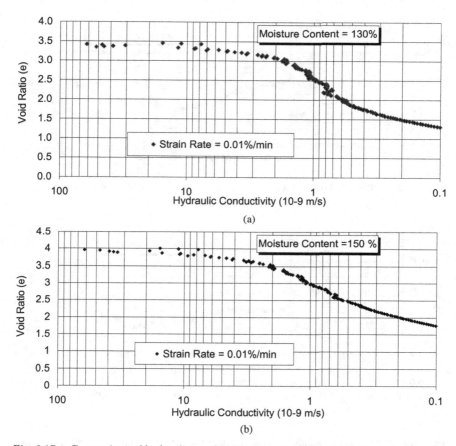

Fig. 8.17. Comparison of hydraulic conductivity measured from various strain rate. (a) $W = 130\%$, (b) $W = 150\%$.

of large strain consolidation is defined as

$$C_F = \frac{k}{\gamma_w(1+e)}\frac{\delta\sigma'}{\delta e} \qquad (8.21)$$

By combining (8.20) and (8.21):

$$C_F = \frac{\delta e \times H^2}{\Delta u_m (1+e_0)^2}\frac{\delta\sigma'}{\delta e} \qquad (8.22)$$

Figures 8.18(a) and 8.18(b) shows the variation C_F with void ratio calculated using (8.22).

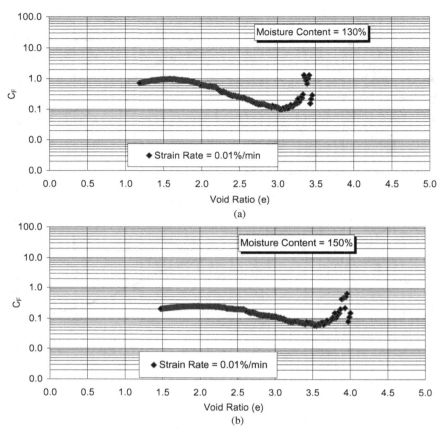

Fig. 8.18. Comparison of void ratio versus coefficient of consolidation (C_F) from various strain rate (a) $W = 130\%$, (b) $W = 150\%$.

8.3. Summary

(i) *Constant Rate of Loading Test*

The CRL test can provide the compression parameters for the ultra-soft slurry-like soil within a short time than by the conventional step loading test. The rate of loading depends on the initial moisture content. The loading rate of $1\,\text{kPa}/800\,\text{s}$ is recommended for soils with moisture contents between 150% and 170%.

The average pore pressure principle is applicable for calculating the effective stress and hydraulic conductivity. In the tests on ultra-soft soil with moisture contents between 150% and 170%, the change in effective stress can be obtained from the average pore pressure by applying an α value of 0.75. Hydraulic conductivity and large strain coefficient of consolidation can be obtained from measurements of volume change, pore pressure, and settlement. These parameters from the CRL test can then be used to predict the magnitude and rate of settlement of ultra-soft soil.

(ii) *Constant Rate of Strain Test*

The CRS test is an alternative approach test to obtain the compression parameters for the ultra-soft clay. However, the selection of a correct rate of strain is important in order to be able to obtain a reasonable set of compression parameters. The strain rate can be estimated using (8.13).

The stabilized excess pore pressure during the test should be less than 30% of the applied pressure, although higher excess pore pressure ratio may be experienced in the early stages of the test due to the viscous effect. Compression parameters such as compression indices in various log cycles C_{c1}^*, C_{c2}^*, and C_{c3}^*, hydraulic conductivity k and large strain coefficient of consolidation C_F can all be obtained from the CRS test. If the selected rate of strain is not correct, the resulting e-log σ' curve for the ultra-soft soil will exhibit pseudo-preconsolidation pressure.

(iii) *Review on All Experimental Tests*

The large-scale consolidometer is only suitable for providing a broad understanding of the deformation behavior of ultra-soft soil. Due to the variation of stress along the column caused by silo effect and wall friction, it is difficult to obtain accurate e-log σ' and e-log k relationships from such tests unless more comprehensive array of pore pressure measurements and void ratio change measurements can be carried out. The small-scale tall column consolidometer also involves large variation of void ratio along the column due to different distance from the drain and is also affected by the friction since D/H ratio is low. Therefore, the void ratio, strain, strain rates, hydraulic gradient, hydraulic conductivity, and effective stresses were only average values.

Samples with high D/H ratio tested in Rowe cell using CRS and CRL are suitable for the measurement of compressibility parameters, the e-log σ'_v and e-log k relationship of ultra-soft soil. The other suitable method to characterize the ultra-soft soil are the sub-sampling technique developed by Mesri *et al.* (1995) together with a controlled gradient tests developed by Sills *et al.* (1984) and the proposed field instrumentation schemes developed by Leroueil *et al.* (1983).

Chapter 9

VERIFICATION OF PROPOSED FORMULAE AND MODELS WITH LABORATORY MEASUREMENTS

A large strain consolidation theory was proposed by Gibson *et al.* in 1967. Schiffman (1980), Schiffman and Cargill (1981), and Znidarcic and Schiffman (1981) updated and supported the theory. Subsequently several researchers have applied this theory using finite difference or finite element solutions to predict very soft clay consolidation under self-weight as well as under additional surcharge. Cargill and Schiffman (1980) developed FSCON 1-I and FSCON 2-I for thick normally consolidated soil, whereas NFSCONSOL was developed by Wu (1994) for very soft soil. Closed-form solutions using dimensionless time factors have been proposed by various researchers to predict the time rate of consolidation in large strains. Gibson *et al.* (1981), Lee and Sills (1981), and Cargill (1982a, b) are among those who proposed various time factor curves. A set of equations with newly proposed compression indices is proposed in the earlier section for the prediction of settlement, whereas the finite difference model to predict time rate of settlement based on Gibson *et al.* (1967) is adopted for ultra-soft soil. It is compared with the existing models proposed by Cargill (1982a, b) and Lee and Sills (1981). The predicted magnitude and rate of settlement were validated with laboratory-measured data for the double drainage condition. The details of the existing and the proposed models, and the corresponding time factor curves will be discussed in this chapter.

9.1. Prediction of Magnitude and Time Rate of Settlement

In order to determine the compression behavior of a clay under additional load, it is necessary to predict both the magnitude as well as the time rate of compression. Some researchers have tried to predict the magnitude and time rate of settlement. Some have been discussed in the earlier chapters and others will be discussed in the following sections.

9.1.1. Magnitude of Settlement

It was reported in Chapter 7 that Carrier III and Beckman's method overestimated the settlement and predicted a decrease in settlement with an increase in the initial moisture content which does not seem logical and differed from the test results obtained in their study. Therefore, their method will not be discussed here again.

In order to validate the reliability of the proposed model, data from step loading compression test are used. Tests were carried out with different boundary conditions such as radial inward, outward, and both. All types of drainage yielded a similar magnitude of settlement and void ratio–effective stress relationships for samples with the same initial moisture content. Therefore, comparison would be confined to one set of data from the radial outward drainage condition. The data were also compared with the e-log σ' curve predicted applying Gibson et al. (1967).

9.1.1.1. Gibson et al. (1981)

Gibson et al. (1967) proposed an e-log σ' relationship.

$$e = (e_i - e_f)\exp(-\lambda\sigma') + e_f \qquad (9.1)$$

$$\lambda = -\frac{1}{\sigma'}\log_e\left[\frac{e - e_f}{e_i - e_f}\right] \qquad (9.2)$$

where e_i and e_f are the initial and final void ratios.

However, if λ is assumed constant the predicted e-log σ' curve may be very different from the measured curve. There are little void ratio changes in low stress range and stress range greater than 100 kPa. Large void ratio changes occurred in the stress range between 10 and 100 kPa, as shown in Figs. 9.1(a) and 9.1(b). Again, predicted results using the constant λ for the Canaveral Harbour Material and Craney Island material reported by

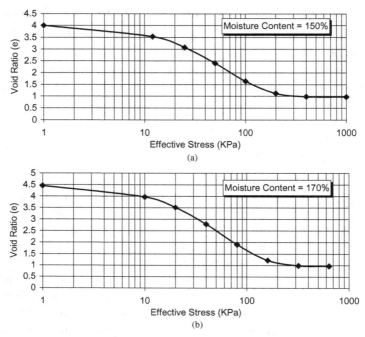

Fig. 9.1. Void ratio–effective stress relationship based on Gibson et al. (1967) if λ is assumed constant: (a) $W = 150\%$; (b) $W = 170\%$.

Cargill (1984) are re-plotted on the log-scale as shown in Figs. 9.2(a) and 9.2(b). The relationship is not linear and is similar to Fig. 9.1.

Gibson et al. (1967) reported the nonlinearity and variability of coefficient λ. The λ values are greater in the low stress range. The value reduces with increasing stress level and becomes more or less constant, as shown in Fig. 9.3. It would be unrealistic to use the constant λ value in the prediction of void ratio change. The predictions are therefore repeated using different λ values at different stress levels. The predicted e-log σ' curves closely match the measured curves as shown in Figs. 9.4(a) and 9.4(b). The variation of λ values applied in the predictions are shown in Figs. 9.5(a) and 9.5(b). Therefore, study of Gibson et al. (1981) requires an advance knowledge of the final void ratio as well as the variation of λ values with stress levels. The method proposed in the present study however requires only the void ratio at the liquid limit.

Prediction of settlement as well as void ratio changes with effective stresses, applying the newly introduced compression indices such as C_{c1}^*,

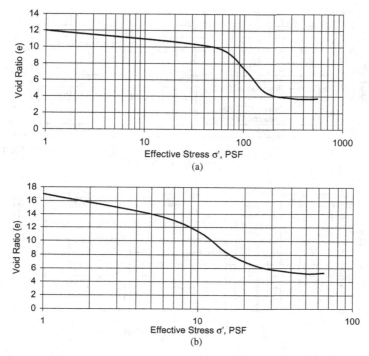

Fig. 9.2. Void ratio–effective stress relationship based on Gibson et al. (1967) when λ is assumed constant. (a) Canaveral Harbour Material; (b) Craney Island Material.

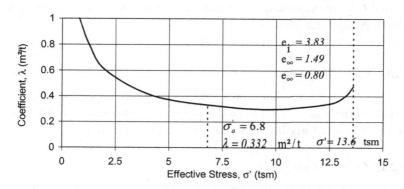

Fig. 9.3. Variation of λ with effective stress (after Gibson et al., 1967).

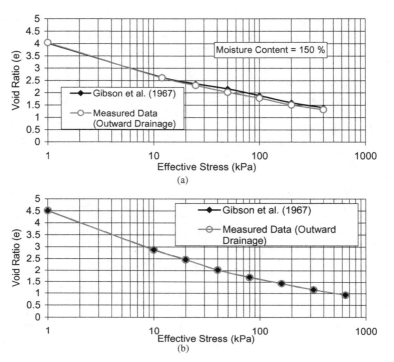

Fig. 9.4. Predicted versus measured void ratio–effective stress relationship after applying varied λ values: (a) $W = 150\%$; (b) $W = 170\%$.

C_{c2}^*, and C_{c3}^* provides good agreement with measured data in the laboratory (Figs. 7.26 and 7.28).

9.1.2. Time Rate of Settlement

Discussions on the methods proposed by previous researchers have been described briefly in Chapter 2. In this section the works proposed by Gibson et al. (1981) and Cargill (1982a, b) will be discussed in detail.

9.1.2.1. Gibson et al. (1981)

Gibson et al. (1967) proposed the following nonlinear finite strain consolidation equation:

$$\pm \left(\frac{\rho_s}{\rho_f} - 1 \right) \frac{\delta}{\delta e} \left[\frac{k(e)}{1+e} \right] \frac{\delta e}{\delta z} + \frac{\delta}{\delta z} \left[\frac{k(e)}{P_f(1+e)} \frac{\delta \sigma'}{\delta e} \frac{\delta e}{\delta z} \right] + \frac{\delta e}{\delta t} = 0 \quad (9.3)$$

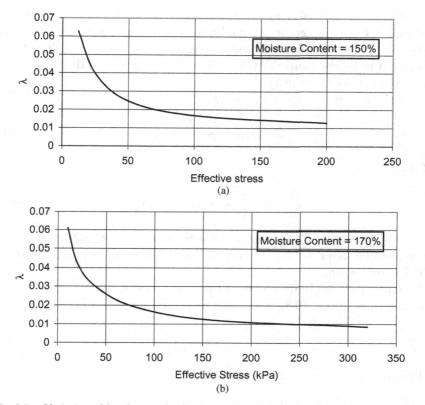

Fig. 9.5. Variation of λ values with effective stress used in the prediction of magnitude of void ratio change applying Gibson et al. (1967) (a) $W = 150\%$; (b) $W = 170\%$.

Since this equation is highly nonlinear due to the nonlinearity of hydraulic conductivity and compressibility, they introduced a finite strain coefficient of consolidation C_F, which is less sensitive to changes in void ratio.

$$C_F = -\frac{k}{P_f}\frac{1}{1+e}\frac{\delta \sigma'}{\delta e} \qquad (9.4)$$

By taking C_F to be constant, a reasonable approximation of void ratio change can be obtained from the following equation:

$$\frac{\delta^2 e}{\delta z^2} \pm (\rho_s - \rho_f)\frac{\delta}{\delta e}\left(\frac{\delta e}{\delta \sigma'}\right)\frac{\delta e}{\delta z} = \frac{1}{C_F}\frac{\delta e}{\delta t} \qquad (9.5)$$

Gibson et al. (1967) stated that despite the simplification achieved by assuming C_F to be constant, nonlinearity still remains by virtue of the presence of the variable coefficient λ as described in Eq. (9.6). When C_F is assumed to be constant, the void ratio–effective stress relationship can be described as shown in (9.1).

Substituting λ from (9.1) into (9.5),

$$\frac{\delta^2 e}{\delta z^2} \pm \lambda(\rho_s - \rho_f)\frac{\delta e}{\delta z} = \frac{1}{C_F}\frac{\delta e}{\delta t} \tag{9.6}$$

Hence,

$$\frac{\delta e}{\delta t} = C_F \left[\frac{\delta^2 e}{dz^2} \pm \lambda(\rho_s - \rho_f)\frac{\delta e}{\delta z} \right] \tag{9.7}$$

Therefore, both the magnitude and time rate of settlement of a compressible layer can be obtained by using (9.1) and (9.7), provided C_F and λ are known.

Gibson et al. (1981) proposed that the effective stress isochrones can be determined from (9.8) where as λ can be obtained from (9.2).

$$\sigma'(z,t) = -\frac{1}{\lambda}\log_e\left[\frac{e(z,t) - e_f}{e_i - e_f}\right] \tag{9.8}$$

The proposed pore pressure isochrones are

$$u(z,t) = p(z,t) - \left\{ \rho_f z + \rho_f \int_0^z e(z,t)\delta z + \rho_f[S(t) + H] \right\} \tag{9.9}$$

where u is the excess pore pressure and p is calculated from (9.10).

$$\sigma' = \sigma - p \tag{9.10}$$

Gibson et al. (1981) proposed curves of time factor versus degree of consolidation for large strains shown in Fig. 9.6. These curves are dependent on "N" values which can be calculated using the relationship,

$$N = \lambda(l)(\rho_s - \rho_f) \tag{9.11}$$

where l is the thickness of the sample or thickness of the layer in material coordinates.

In order to predict the settlement using Gibson et al. (1981), λ and N values at various stages must be known.

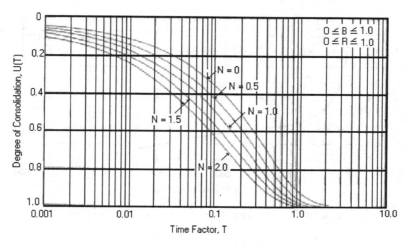

Fig. 9.6. Time factor curves proposed by Gibson *et al.* (1981).

9.1.2.2. Cargill (1982)

Cargill in 1982 developed similar time factor curves following the approach by Gibson *et al.* (1967). The time factor curves applying the large strain theory for underconsolidated and normally consolidated soil of single drainage and double drainage conditions are shown in Fig. 9.15. The time factor can be calculated using the following equation provided the coefficient of large strain consolidation C_F is known:

$$T_F = \frac{C_F t}{z^2} \tag{9.12}$$

where z is the drainage length or layer thickness.

The degree of settlement is again largely dependent upon the coefficient "N" which is derived from the coefficient λ and thickness of the layer. As such, λ and N values are required to be known in advance for prediction. λ can be calculated from initial and final void ratios for various stress ranges using Eq. (9.2) and N can be calculated using Eq. (9.11). The N values were found to range between 0 and 0.9, as shown in Fig. 9.7. Since the difference between the time factor curves for $N = 0$ and $N = 2$ provided by Cargill (1984) for double drainage are very close to each other, the time factor curve for $N = 2$ is used in the comparison. The predictions of time rates of settlement were carried out using time factor curves for double drainage normally consolidated condition. The self-weight consolidated stage was ignored. The measured and predicted time rates of

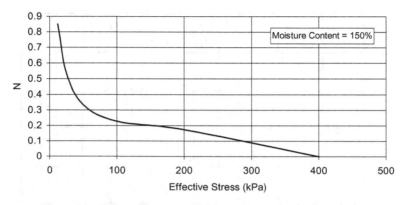

Fig. 9.7. Variation of N with effective stress ($W = 150\%$).

settlement using Cargill time factor curve with $N = 2$ are shown in Figs. 9.8 and 9.9. It can be seen that except for the 10 kPa loading step in the 170% moisture content test, all other cases produce results that are reasonably in good agreement with the measured time settlement curves. However, the agreement was not very good in some of the tests with 150% moisture content.

9.1.2.3. Lee and Sills (1981)

Solutions for the consolidation of a soil stratum including self-weight effects and large strains were also proposed by Lee and Sills (1981). Their method was also based on the governing equation proposed by Gibson et al. (1967). The magnitude and time rate of deformation were determined from the stress–strain and hydraulic conductivity considerations, respectively. For the self-weight consolidation, they proposed approximate equations for single and double drainage conditions.

For double drainage,

$$U_s = \sqrt{\frac{\pi}{4}T}, \quad \text{for small } T \text{ values} \tag{9.13}$$

$$U_s = 1 - \frac{8}{\pi^2}\exp(-\pi^2 T), \quad \text{for large } T \text{ values} \tag{9.14}$$

where U_s is degree of settlement.

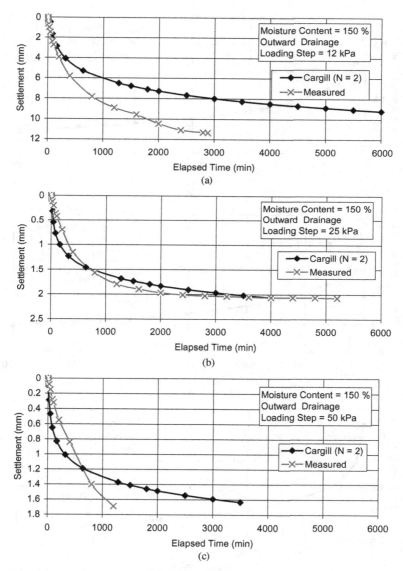

Fig. 9.8. Measured and predicted time settlement curves: Loading step (a) 12 kPa (b) 25 kPa (c) 50 kPa (d) 100 kPa (e) 200 kPa (f) 400 kPa.

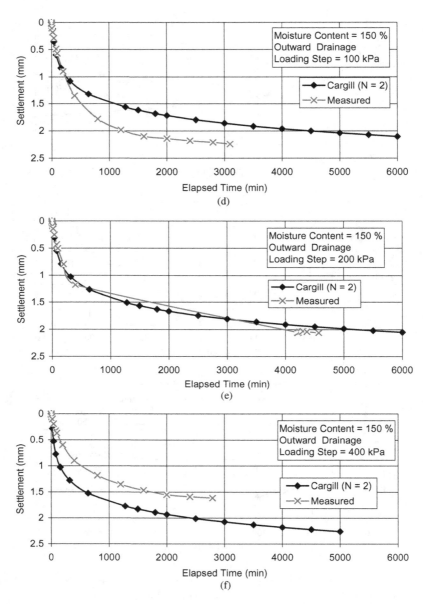

Fig. 9.8. (*Continued*).

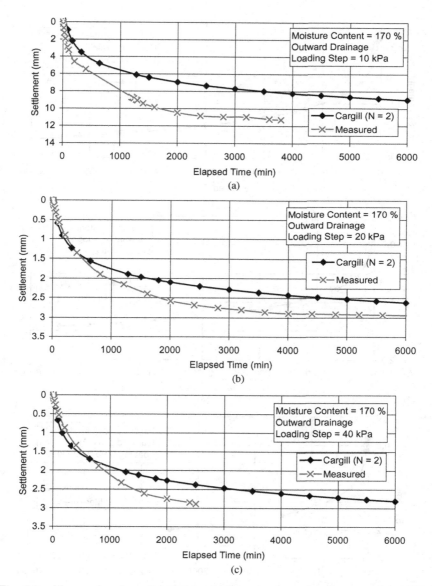

Fig. 9.9. Measured and predicted time settlement curves: Loading step (a) 10 kPa (b) 20 kPa (c) 40 kPa (d) 80 kPa (e) 160 kPa (f) 320 kPa (g) 640 kPa.

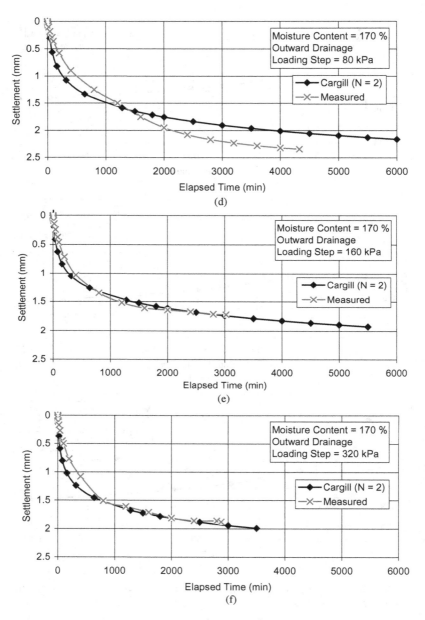

Fig. 9.9. (Continued).

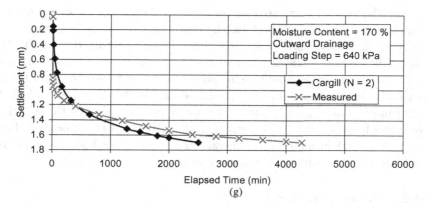

Fig. 9.9. (*Continued*).

For single drainage,

$$U_s \approx 2T \qquad \text{for small } T \text{ values} \qquad (9.15)$$

$$U_s = 1 - \frac{2}{\pi^3}\sum_n \frac{1}{m^2}\exp(-m^2\pi^2 T) \quad \text{for large } T \text{ values} \qquad (9.16)$$

After the soil has attained an effective stress equivalent to its self-weight, Lee and Sills switch to new sets of equations for consolidation assuming linear effective stress–void ratio relationship and equal change of void ratio in the stratum. Their new sets of equations for the time rate of settlement are largely dependent upon the length of drainage path as in Terzaghi's theory.

For double drainage,

$$U_s = 1 - \frac{8}{\pi^2}\sum_n \frac{1}{m^2}\exp(-m^2\pi^2 T) \qquad (9.17)$$

$$m = (2n+1), \quad n = 0, 1, 2 \ldots$$

For single drainage,

$$U_s = 1 - \frac{2}{\pi^2}\sum_n \frac{1}{m^2}\exp(-m^2\pi^2 T) \qquad (9.18)$$

where

$$m = \frac{1}{2}(2n+1), n = 0, 1, 2 \ldots$$

For the condition covered in this study, the second set of equations is assumed to be valid. Therefore, the self-weight consolidation stage was neglected. In both equations the time factor T can be obtained using (9.12). The predicted time settlement curves are compared with the measured data for the double drainage condition in two different moisture contents of 150% and 170%. It was found that the predicted curves of Lee and Sills yielded better match than Cargill's (1982) time factor curves including those in the low stress level of 10 or 12 kPa stress (Figs. 9.10 and 9.11). Moreover, Lee and Sills (1981) proposed only one set of equations for both double and single drainage conditions. No additional parameter is required.

9.1.2.4. Proposed Model

The proposed model basically followed the approach by Gibson et al. (1967). Lee and Sills (1981) proposed the following equation:

$$\frac{\delta e}{\delta t} = C_F \frac{\delta^2 e}{\delta z^2} \tag{9.19}$$

This simplified equation was transformed into an implicit finite difference equation as follows:

$$\frac{e_{i,t+\Delta t} - e_{i,t}}{\Delta t} = C_F \frac{\left(\frac{e_{i-1}-2e_i+e_{i+1}}{(\Delta z)^2}\right)_t + \left(\frac{e_{i-1}-2e_i+e_{i+1}}{(\Delta z)^2}\right)_{t+\Delta t}}{2} \tag{9.20a}$$

Therefore,

$$e_{i,t+\Delta t} = \frac{(e_{i-1} + e_{i+1})_{t+\Delta t} + e_{i,t}}{2(1+\frac{1}{\beta})} \tag{9.20b}$$

$$e_{i,t} = \left[e_{i-1} - 2\left(1 - \frac{1}{\beta}\right)e_i + e_{i+1}\right]_t \tag{9.20c}$$

$$\beta = \frac{C_F \Delta t}{(\Delta z)^2} \tag{9.20d}$$

In the proposed model, the magnitude of settlement was calculated applying Eqs. (7.16) and (7.17). The input data for the equations are initial void ratio e_i, initial thickness H_0, final stress σ'_f, and various compression indices C^*_{c1}, C^*_{c2} and C^*_{c3}. In order to obtain an accurate predicted settlement, layers can be subdivided and input data for sublayers can be varied. After obtaining the magnitude of settlement, final void ratio can be calculated. Input data of initial void ratios along the depth of

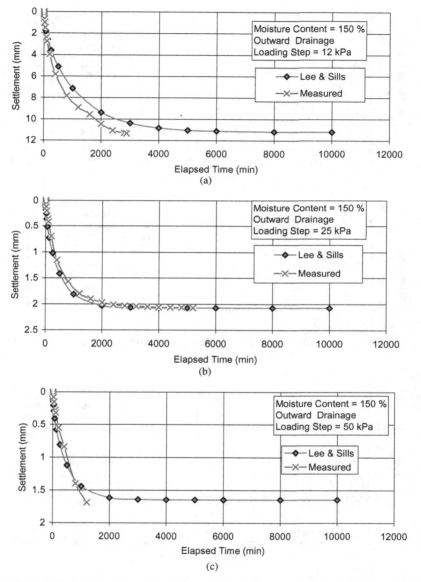

Fig. 9.10. Measured and predicted time settlement curves: Loading step (a) 12 kPa (b) 25 kPa (c) 50 kPa (d) 100 kPa (e) 200 kPa (f) 400 kPa.

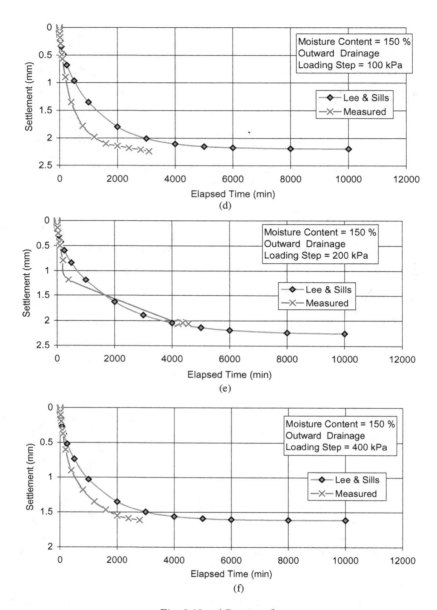

Fig. 9.10. (*Continued*).

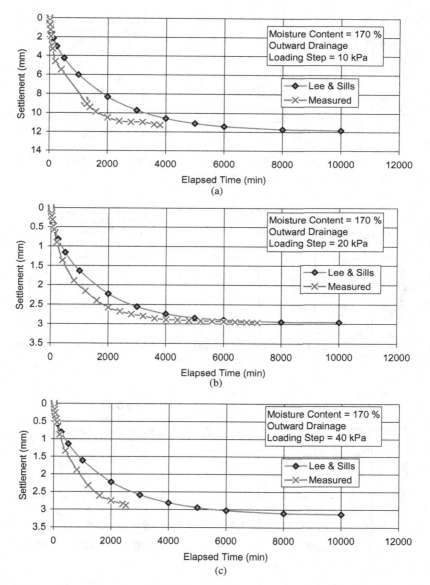

Fig. 9.11. Measured and predicted time settlement curves: Loading step (a) 10 kPa (b) 20 kPa (c) 40 kPa (d) 80 kPa (e) 160 kPa (f) 320 kPa (g) 640 kPa.

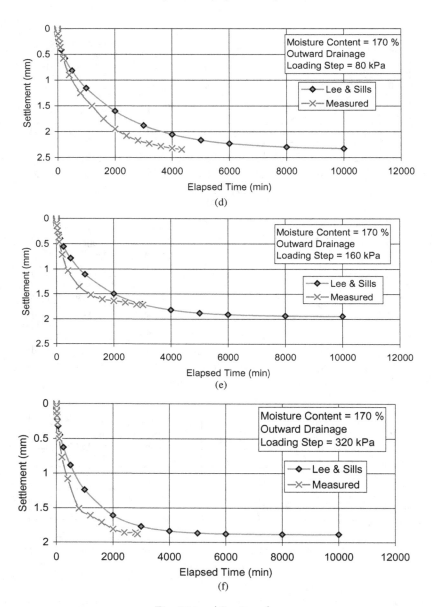

Fig. 9.11. (*Continued*).

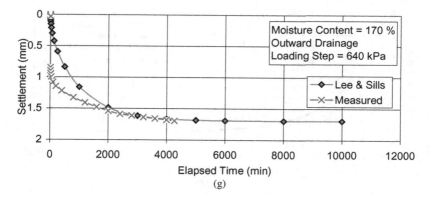

Fig. 9.11. (*Continued*).

profile can be varied. Final void ratios are put at the drainage boundaries and kept constant throughout the calculations. Two types of boundary conditions such as single and double drainage can be modeled by changing the Δz accordingly and by inputting relevant final void ratios at relevant boundaries. The vertical drain condition can be modeled as double drainage condition and the same C_F value can be used since ultra-soft soil is isotropic in compression as well as hydraulic conductivity. Stage construction can also be modeled in the proposed model in which final void ratios for the relevant final stresses in the stages were calculated first and these final void ratios became the input data. After that, time rates of settlement for various stages were calculated with the given time duration and the void ratios at time "t" (e_t) was obtained. This void ratio again became the initial void ratio for the next step. This procedure was repeated for all the stages. This program was written on the spreadsheet by the author and all the predictions presented in the chapter were performed by the author. Figures 9.12 and 9.13 show a comparison between the predicted time rate of settlement for two different moisture contents (150% and 170%) under various load steps. It can be seen that the proposed solution predicted very closely with the measured data at all stress levels.

9.2. Comparison of Various Large Strain Existing Models with Proposed Time Factor Curve

The time factor curves produced from the proposed model were compared with the curves proposed by others such as Cargill (1984) and Lee and Sills

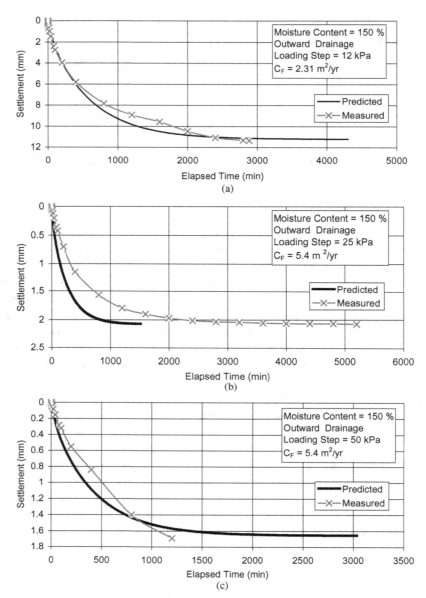

Fig. 9.12. Measured and predicted time settlement curves applying present study: Loading step (a) 12 kPa (b) 25 kPa (c) 50 kPa (d) 100 kPa (e) 200 kPa (f) 400 kPa.

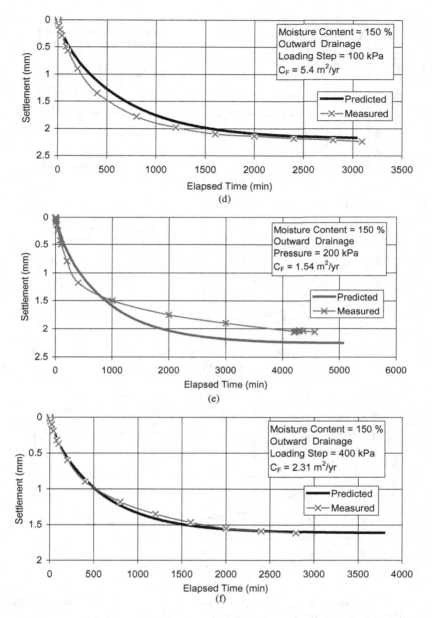

Fig. 9.12. (*Continued*).

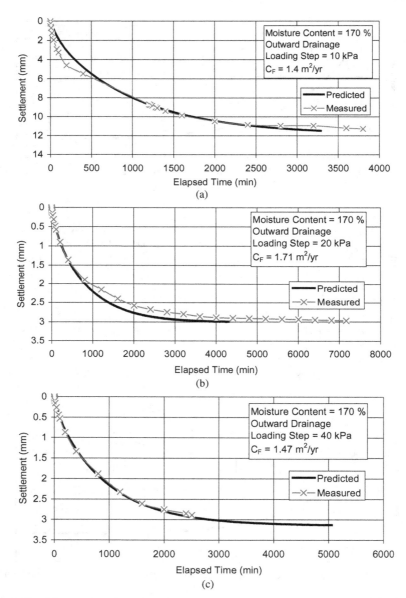

Fig. 9.13. Measured and predicted time settlement curves: Loading step (a) 10 kPa (b) 20 kPa (c) 40 kPa (d) 80 kPa (e) 160 kPa (f) 320 kPa (g) 640 kPa.

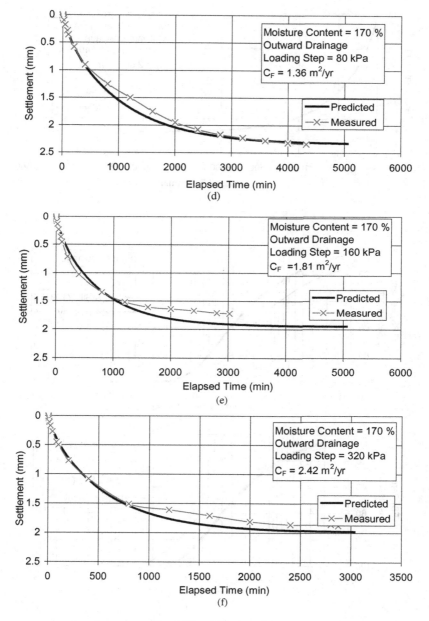

Fig. 9.13. (*Continued*).

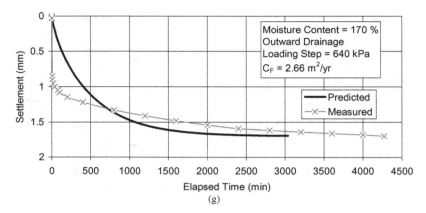

Fig. 9.13. (*Continued*).

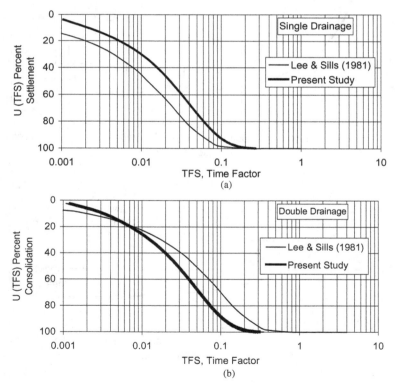

Fig. 9.14. Comparison of time factor curves from the present study and from the study of Lee and Sills (1981): (a) single drainage (b) double drainage.

(1981). The predictions from the model described in the earlier section are in close agreement with those by Lee and Sills for normally consolidated conditions with both single and double drainage as shown in Figs. 9.14(a) and 9.14(b). The proposed model predicts a slightly slower rate than that of Lee and Sills (1981) in single drainage and a faster rate in double drainage. This could be due to the differences in the methods of prediction. Lee and Sills (1981) predicted a degree of consolidation based on exponential function, whereas the proposed model is solely based on the rate of void ratio change.

The time factor curves from the proposed model were compared with those by Cargill (1984), as shown in Figs. 9.15(a) and 9.15(b). The

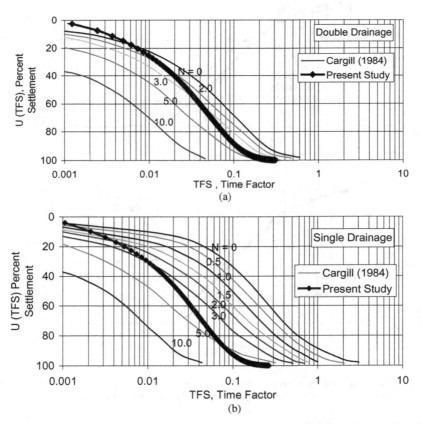

Fig. 9.15. Comparison of time factor curves from the present study and from Cargill (1984): (a) double drainage (b) single drainage.

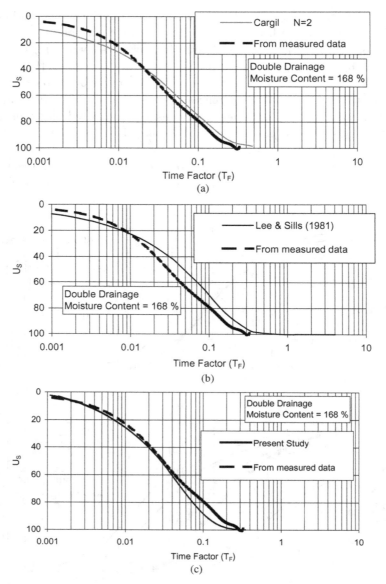

Fig. 9.16. Time factor curves, double drainage condition compared with time factor curves from measured data: (a) Cargill, 1984 (b) Lee and Sills, 1981 (c) proposed model.

agreement was not good. Among all the methods, Lee and Sills' (1981) method is straightforward and requires only C_F values and a single time factor curve. Cargill's method requires prior knowledge of relevant N values for various time factor curves. The proposed model requires only C_F value just as in Lee and Sills (1981) and has only a single time factor curve for each drainage condition. In addition, the prediction from the proposed model agrees better with the measured data in both low and high stress ranges (Figs. 9.12 and 9.13). The time factor curves produced from the proposed model were compared with the time factor curves produced from the measured data described in Chapter 7. It was found that time factor curves from the proposed model closely match with those from the measured data (Fig. 9.16(c)). The time factor curve from Cargill (1984) with $N = 2$ closely matches that from the measured data.

9.3. Summary

Most of the existing methods of predicting magnitude of settlement and time rate of settlement basically follow the large strain theory by Gibson et al. (1967). Carrier III and Beckman (1984) introduced new parameters, whereas Tan et al. (1988) proposed a single nonlinear e-log σ' curves based on experimental measurements.

For the prediction of magnitude of settlement, discussions and comparisons were made only with those from Gibson et al. (1967) and Carrier III and Beckman (1984). The method by Carrier III and Beckman (1984) requires power functions which are related to physical parameters. However, their method predicted higher settlement for low initial moisture content soil and lower settlement for high initial moisture content soil which contradicted with the measurements from the experimental results discussed in the earlier sections. The method by Gibson et al. (1967) requires λ values at various stress ranges to be able to predict the magnitude accurately, whereas the proposed method requires only compression indices (C_{c1}^*, C_{c2}^*, and C_{c3}^*) which are related to the liquid limit. The proposed method closely predicts the magnitude of settlement from the measurements.

Comparisons of time rate of settlement predictions were made with the time factor curves produced by Cargill (1984) and Lee and Sills (1981). The method by Cargill (1984) requires the prior knowledge of the parameter N to select the appropriate time factor curve. The predicted curves did not agree with the measured curves in the low stress range. Lee and Sills (1981) proposed a single set of equations for each type of drainage condition and requires only a large strain coefficient of consolidation C_F value. The predicted curves agree reasonably well with the measurements over most of the stress ranges.

Based on the results from the experimental tests, an effective stress model with trilinear semi-log e-log σ'_v relationship to reflect the various compression indices in the different stages of soil such as ultra-soft soil stage and soil stage was proposed. The time rate of settlement model was written with an implicit finite difference equation applying Gibson et al. (1967) large strain consolidation theory. Time factor curves for both single and double drainage conditions have also been presented for conventional manual computation. The only parameter required is the large strain coefficient of consolidation (C_F) which can either be obtained from the test data or estimated from the liquid limit and void ratio by using Eqs. (7.15) and (8.21). In order to be able to predict the time rate of settlement in the various steps, the varied C_F values in the different stress levels were used although C_F were kept constant for each step. The proposed model closely predicts the time rate of settlement in both the low and high stress ranges. The model can be used for nonhomogenous ultra-soft soil with varying void ratios along the profile.

Chapter 10

CASE STUDY

A case study of the compression behavior of ultra-soft soil was carried out at the reclamation project in Changi East, Singapore. The reclaimed project is situated at the offshore area in the eastern part of Singapore. An ultra-soft slurry-like soil of about 10–15 m thickness with very high water content was contained in a bunded area. The material inside the silt pond had virtually no shear strength and water content as high as 250%. The material was undergoing self-weight consolidation. The settlement and pore pressure behavior after filling of the silt pond was monitored in two designated areas. The proposed model for predicting magnitude and time rate of settlement was validated using the monitored results from the test areas.

10.1. Description of the Silt Pond

During the reclamation works for Changi Airport, between 1972 and 1978 the borrowing of sand created a pit in the seabed to an elevation of about −22 m CD (Chart Datum). In 1986, a containment sand bund was constructed around this borrow pit. Silt and clay washings from sand quarrying activities inland were transported in pipelines with high water content and discharged into the bunded area (Fig. 10.1). The pond was locally named as silt pond although majority of fines are clay. The silt pond was trapezoidal in shape and had an area of about 180 hectares. It measured about 2000 m in length and 750 m and 1050 m in width at the two ends.

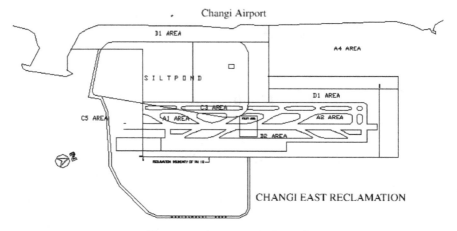

Fig. 10.1. Location of silt pond.

10.1.1. *Soil Condition in the Silt Pond*

A subsurface investigation was conducted in 1992. This included 82 gamma-ray density probes and 20 boreholes with sampling and *in situ* vane tests. Since the soil was very soft, the sampling was carried out with a twist sampler (Fig. 10.2). Further subsurface investigation work including boreholes, field vane tests and gamma-ray density probes resumed again in 1994.

The elevation of the top of the sediment varied from 0 to -5 m CD with an average of about -3 to -4 m CD. The top of the slurry was taken as the elevation at which the density was greater than $1.1\,\mathrm{Mg/m^3}$. The elevation of the bottom of the slurry varied from 0 to -22 m CD with an average -12 to -18 m CD. The thickness of sediment varied from 1 to about 20 m with an average of about 10 to 15 m. Various types of soil investigation locations in the silt pond are shown in Fig. 10.3. Contour of depth to various density layers are shown in Figs. 10.4 and 10.5. A typical density profile determined from Gamma–Gamma logging is shown in Fig. 10.6.

10.1.2. *Characterization of the Silt Pond Slurry Prior to Reclamation*

The silt pond material prior to reclamation consisted of low to high plasticity clay of very high water content and low bulk density (Fig. 10.9).

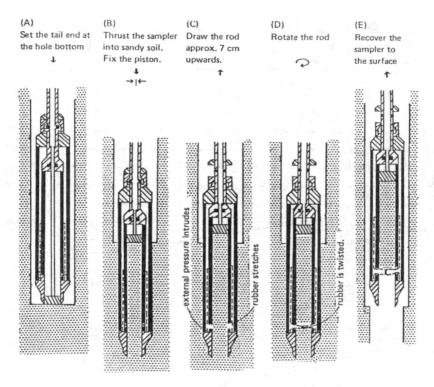

Fig. 10.2. Twist sampler.

The water content was as high as 250% with an average of 170%. The material was still undergoing self-weight consolidation. The density of the top part of clay was only slightly above that of salt water density. The mean grain size of clay (D_{50}) was smaller than 3 μm and about 80% of the particles were smaller than 63 μm (Fig. 10.7). The strength of the slurry was as low as 0.5 kPa at the surface and increases with depth at only 0.2 kPa/m. A summary of the properties of the silt pond material is shown in Table 10.1, and their classification is shown in Fig. 10.8. The properties versus depth are shown in Fig. 10.9. It can be seen in the figures that the material had a very high moisture content. The preconsolidation pressures were much lower than the overburden pressure which confirmed that the material was underconsolidated and still undergoing self-weight consolidation. Although underconsolidation is not an indication of transition point between slurry and soil, however it indicates it is still undergoing self-weight consolidation.

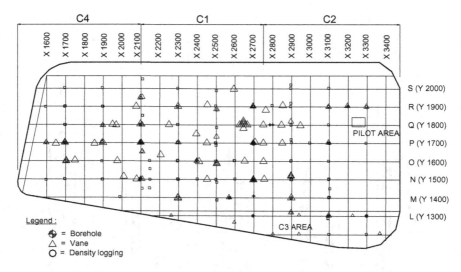

Fig. 10.3. Locations of density logging, boreholes, and field vane.

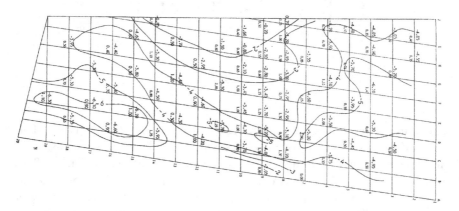

Fig. 10.4. Contour map showing depth to density $1.3\,\text{g/cm}^3$.

The liquidity index decreased with depth. This supports the findings by Been and Sills (1981) on self-weight consolidation which suggested that consolidation begins from the bottom part of the soil rather than from the surface. Results of density logging also showed that the density was increasing with depth. This confirmed that consolidation commenced from the bottom rather than from the surface. Since the material was

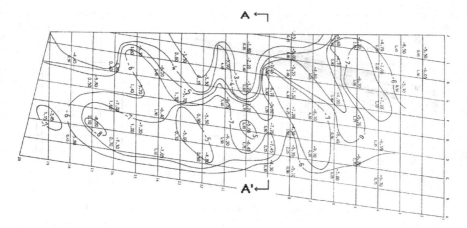

Fig. 10.5. Contour map showing depth to density $1.5\,\text{g/cm}^3$.

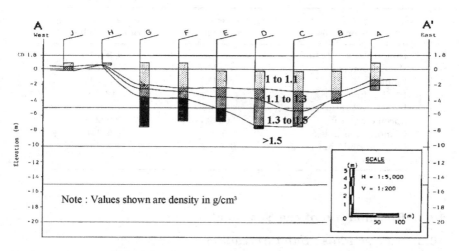

Fig. 10.6. Density profile along Section A–A'.

in an underconsolidated state, its void ratio was much larger than the transition void ratio which was expected to be more or less equal to the void ratio at the liquid limit. The liquid limit of the material was between 60% and 115%, and the corresponding void ratios are 1.74–3.08, respectively.

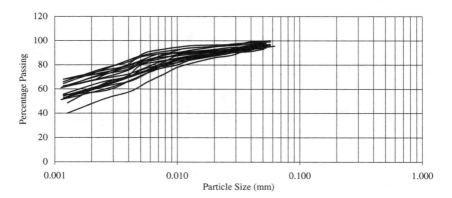

Fig. 10.7. Grain size distribution curve of silt pond materials.

Table 10.1. Soil properties of silt pond clay prior to reclamation and after sand spreading.

Properties	Prior to reclamation range of values	After sand spreading range of values
Bulk density (Mg/m^3)	1.25–1.6	1.3–1.9
Water content (%)	75–170	40–150
Liquid limit (%)	60–115	50–105
Plastic limit (%)	22–45	18–30
Initial void ratio	2–4.5	2–3.5
Specific gravity	2.66–2.7	2.5–2.7
Compression index	0.5–1.7	0.6–1.4
Overconsolidation Ratio	0.2–1.2	0.2–1.0
$\Delta Cu/\Delta p'$	0.078	0.094

10.2. Reclamation of the Silt Pond

10.2.1. *First Phase Sand Spreading*

The material at the surface of the silt pond was extremely soft and had virtually no shear strength. Direct hydraulic filling or dumping of sand on this ultra-soft foundation was not possible. A sand spreading method was therefore proposed. In this method sand was pumped through pipelines with high water content and deposited in a loose form with low density. The spreader used in the sand spreading work is shown in Fig. 10.10. The photographic features of silt pond prior to reclamation and sand spreading in progress are shown in Figs. 10.11(a) and 10.11(b) respectively.

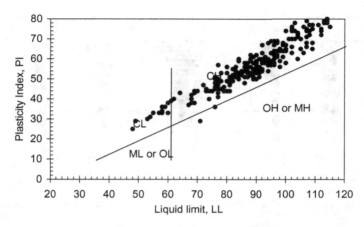

Fig. 10.8. Silt pond material on classification chart.

The sequence of sand spreading in the first phase in the silt pond consisted of evenly spreading sand in 20 cm lifts until the total sand thickness deposited was 2 m. A certain rest period was required after each 20 cm lift to ensure stability of the foundation.

A hydrographic survey was carried out after each lift to verify the thickness of the deposited sand. Hydrographic survey profiles after each phase of sand spreading are shown in Fig. 10.12. The sand spreading in this phase was carried out to an elevation of about −1 to 0 m CD.

10.2.2. Site Investigation During and After First Phase of Sand Spreading

In order to determine the build up of sand on top of the slurry and to investigate the improvement of the slurry, interim boreholes, field vane shear tests, and density loggings were carried out. The build up of sand were found to be up to 5.2 m. The increase in density at the bottom part of slurry but not at the top of the soil were noted at several places. Figure 10.13 shows a comparison between typical density logging results before and after the sand spreading. It can be seen that the bottom portion of the slurry has reached the soil stage whilst the top portion was still in the slurry stage.

This supports Been and Sills' (1981) assumption that the process of self-weight consolidation begins from the bottom part due to a higher

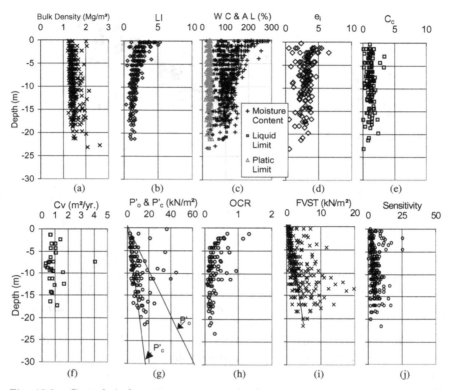

Fig. 10.9. Geotechnical parameters versus depth prior to sand spreading: (a) bulk density; (b) liquidity index; (c) water content and Atterberg limit; (d) initial void ratio; (e) compression index; (f) coefficient of consolidation; (g) overburden and preconsolidation pressure; (h) overconsolidation ratio; (i) shear strength from field vane test; (j) sensitivity from field vane test.

overburden pressure at the bottom. It was also obvious that the build up of density started from the bottom of the slurry. Characteristics of the material inside the silt pond after sand spreading to about 0 m CD are shown in Figs. 10.14(a)–10.14(j).

It can be seen in the figures that the bulk densities of soil increased to about 1.3 to 1.5 Mg/m^3. However, from the liquidity index the moisture content in the top 10 m was still above the liquid limit. The void ratio reduced to about 3 to 4 within the top 10 m and to about 2 or slightly greater at depths below 10 m. The preconsolidation pressures of most cases were still below the effective overburden pressure (even with the exclusion of the additional load from the new sand fill). This indicated that the soil

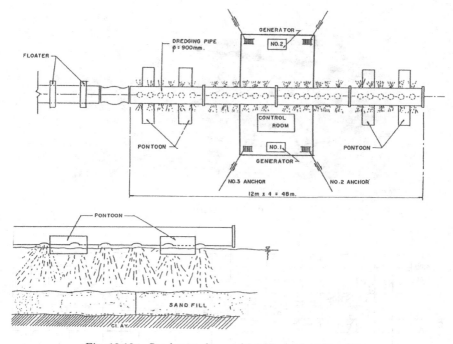

Fig. 10.10. Sand spreader used in silt pond reclamation.

(a)

Fig. 10.11. (a) Silt pond prior to reclamation; (b) Sand spreading in silt pond in progress.

(b)

Fig. 10.11. (*Continued*).

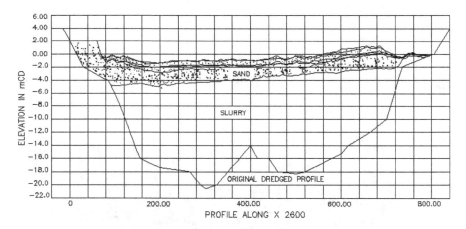

Fig. 10.12. Hydrographic survey profile after each phase of spreading.

was underconsolidated. The overconsolidation ratio of material was below unity. The undrained shear strengths measured by field vane tests were still below 5 kPa with an average of 2 kPa up to 7 m depth and 5–7 kPa below 7 m depth. Based on the study of Leroueil (1999), the measured undrained shear strengths are reasonable when the water content is close to liquid limit. The summary of soil properties in silt pond after the first phase of sand spreading is shown in Table 10.1. It could be seen that only a slight improvement of the parameters is evident.

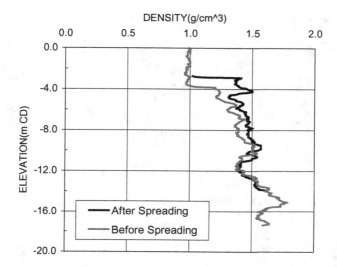

Fig. 10.13. Comparison of density profile after first phase of sand spreading.

10.2.3. Failure During Sand Spreading

Although extreme care was taken during the sand spreading, a failure occurred at one location. Due to the unevenness of the original pond bed, sand formation of varying thickness was deposited. The process of sand filling attempted to achieve in a uniform thickness of each sand layer; however investigation after the first phase of sand spreading showed sand formations varying from 1 to 7 m thickness. The resulting difference in additional load and differential fill pressure caused the soft slurry to be displaced toward the locations with low fill pressure and the slurry burst through the thin sand cap near the center of the silt pond. Evidence of this type of failure was confirmed in the post-failure investigation using wash-boring and marine Dutch Cone penetration tests. Heaved up clay was found in the silt pond over an area of 700 m by 500 m and the clay had heaved up as high as +2 m CD. The condition of the silt pond slurry after failures can be characterized as a very soft viscous liquid. The failure area is shown in Fig. 10.15.

10.2.4. Remedial Measures

The heaved up mud at the failure location were removed by a high capacity submersible mud pump. After the partial removal of the mud, geotextiles of

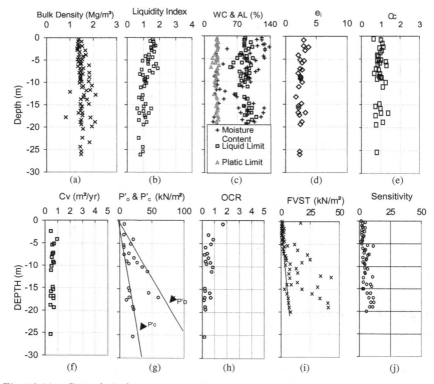

Fig. 10.14. Geotechnical parameters versus depth after first phase of sand spreading: (a) bulk density; (b) liquidity index; (c) water content and Atterberg limits; (d) void ratio after sand spreading; (e) compression index; (f) coefficient of consolidation; (g) overburden and preconsolidation pressure; (h) overconsolidation ratio; (i) field vane shear strength; and (j) sensitivity from vane shear test.

strength 150 kN/m by 150 kN/m with dimensions of 700 m by 900 m were placed in two layers over the failure location and the surrounding areas prior to further sand spreading. Details of the design and construction method are discussed by Na et al. (1998). The layout of the geofabric placement is shown in Fig. 10.15. Photographic feature of geofabric laying in progress is shown in Figs. 10.16(a) and 10.16(b).

10.2.5. Second Phase of Sand Spreading

Following the placement of the geotextile in the silt pond, the second phase of sand spreading resumed. To enable sand to be spread up to +4 m CD

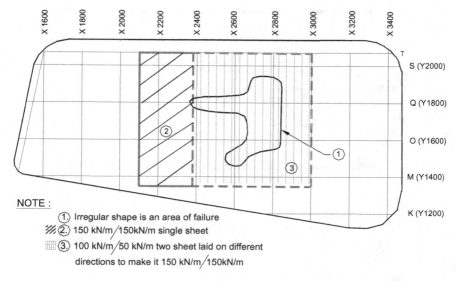

Fig. 10.15. Layout of geofabric placement in the center of silt pond.

with the spreading barge, the surrounding containment bund elevation was raised to +6 m CD and the water level in the silt pond was subsequently raised to +5.5 m CD. Hydrographic surveys were carried out following the completion of this phase of sand spreading to verify the actual build up of sand over the geofabric. This method of sand spreading was proceeded until the elevation of the silt pond reached +3.8 to +4 m CD. It took 8 months to reach the elevation of +4 m CD. Altogether, it took 21 months from the beginning of the first phase of sand spreading including the placement of geotextile.

10.2.6. *Intermediate Mini-Cone Penetration Tests*

When the fill level of sand in the silt pond reached about +4 m CD, 100 numbers of marine mini-cone penetration tests were carried out using a 3 ton mini-seacalf mounted on a floating pontoon (Fig. 10.17). The purposes of these tests were to determine the total thickness of the spread sand, to verify the success of sand spreading, and to detect the possiblity of any further local failures. The mini-cone penetration tests indicated a trend of a linear increase of cone resistance with depth for the sand found above the geofabric. This trend confirmed the formation of the spread sand in

Fig. 10.16. (a) Geofabric laying over failure area. (b) Geofabric laying in progress.

sufficient thickness over the geotextile. In addition to this, the mini-CPT also picked up lenses of sand which seemed to have intruded into the slurry in the first phase of sand spreading. The typical cone resistance profiles measured by the mini-cone is shown in Fig. 10.18. The thickness of the sand profile determined by CPTs after reaching +4 m CD is shown in Figs. 10.19 and 10.20. It can be seen in these figures that there was a build up of 8–11 m thickness of sand on top of the silt pond material.

10.2.7. *Characterization of Soil after Sand Spreading*

After the sand was filled up to the +4 m CD level, several interim boreholes were carried out to investigate the improvement of the slurry. An increase in bulk density, reduction in void ratio and moisture content, and a slight

Case Study

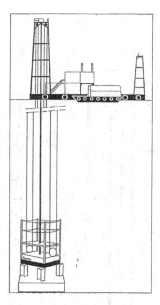

Fig. 10.17.   Mini-seacalf.

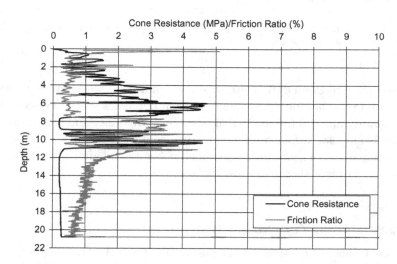

Fig. 10.18.   Typical cone resistance profile after sand spreading.

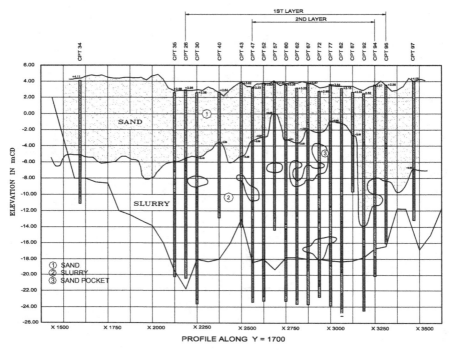

Fig. 10.19. Cross-sectional profile showing build up of sand after sand spreading (interpreted from mini-CPT).

increase in preconsolidation pressure and undrained shear strength were registered from laboratory results of interim boreholes. It can be seen that the liquidity index of the top 7–10 m was still above unity and void ratio was above 2. Therefore, the material up to 10 m depth was still in a slurry stage. The comparison of geotechnical parameters before and after reclamation to +4 m CD levels is shown in Figs. 10.21(a)–10.21(f).

10.2.8. Lowering of Ground Water Level

The next phase of construction in the silt pond was the lowering of ground water in the silt pond to +3 m CD so that the prefabricated vertical drains could be installed at an elevation of +4 m CD. This operation was carried out by opening the control sluice pipes that were installed in the silt pond sand bund prior to the resumption of sand spreading activities. Internal

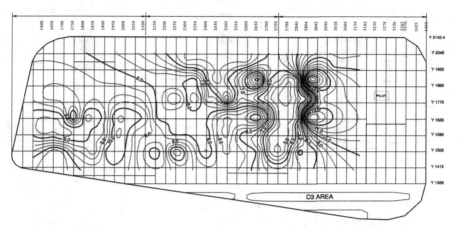

Fig. 10.20. Thickness of built up sand after sand spreading interpreted from mini-CPT data.

dewatering trenches together with a pumping scheme were provided to accelerate the lowering of groundwater level.

10.2.9. *Installation of Vertical Drains and Surcharges*

After the sand filling, a loose 4–10 m thick sand layer overlies the thick slurry within the bund. Due to the slow process of filling combined with slow dewatering, additional load on the slurry was gradually imposed. However, the pore pressure in the slurry was still high since only a limited quantity of pore water was drained out. In order to release the pressure, vertical drains were installed at 2 m square spacing. Some viscous liquid slurry first came out through the annulus between mandrel hole and vertical drain during the installation. Thereafter clear water released through the vertical drain. Subsequently, more sand was placed stage by stage up to +9 m CD all over the silt pond.

10.3. **Pilot Embankment and Other Test Areas**

In order to investigate the deformation and pore pressure dissipation of the slurry-like soils, soil instrument clusters were installed after filling to +4 m CD. Instruments installed in a typical cluster included surface settlement

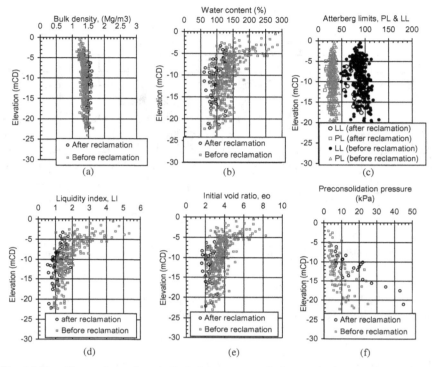

Fig. 10.21. Comparison of geotechnical parameters before and after sand spreading: (a) bulk density; (b) water content; (c) Atterberg limits; (d) liquidity index; (e) initial void ratio; (f) preconsolidation pressure.

plates, deep settlement gauges, pneumatic piezometers, electric vibrating-wire piezometers, open-type piezometers and water stand-pipes. The deep settlement gauges and piezometers were installed at various elevations in the soil. The arrangements of each cluster in plan and elevation are shown in Figs. 10.22–10.24. In these clusters, specially modified piezometers with protected guard cells were installed, as shown in Figs. 10.25 and 10.26. With this type of installation, the settlement of the piezometer tip was measured through surveying the level at the top of the extended pipe.

10.4. Deformation Behavior of the Silt Pond Slurry

Instruments in the pilot area registered a total settlement of 2.5 m within a one-year period. This settlement is equivalent to a strain of 30%. This

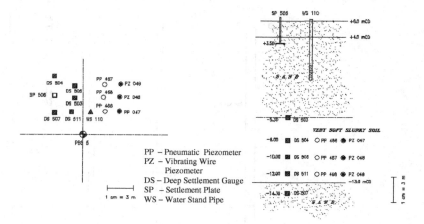

Fig. 10.22. Instrument layout and installed elevations at silt pond pilot area.

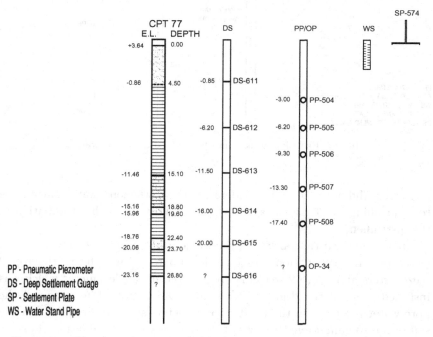

Fig. 10.23. Elevation of instruments installed at an area with largest settlement.

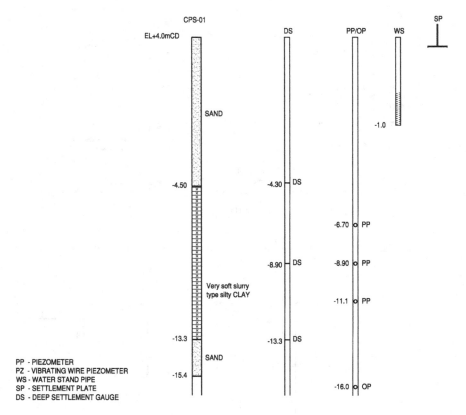

Fig. 10.24. Elevation of instruments installed at the no-drain area.

settlement did not take into consideration the settlement that occurred due to sand filling up to an elevation of +4 m CD prior to the installation of the instruments.

It was noticed that although the foundation soil was settling, no pore water pressure dissipation was recorded up to 11 months in the top 2 piezometers and up to 2 weeks in the bottom-most piezometer which was installed close to the drainage boundary. Only after 11 months did the pore water pressures in the top two piezometers start to dissipate and the soil began to gain strength. Actually there was a small reduction of excess pore pressure during the initial 11-month period. This was not due to pore pressure dissipation, but rather due to the submergence of the sand fill. The observation of no effective stress and strength gain despite the occurrence

Case Study

Fig. 10.25. Piezometers with protected guard shell.

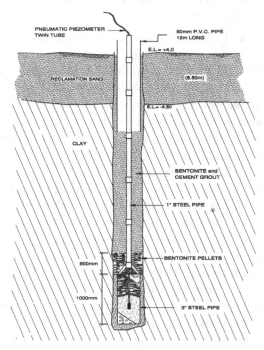

Fig. 10.26. Detail of specially protected pneumatic piezometer with extended protection pipe.

of large settlement had been confirmed by various *in situ* and laboratory tests (Bo *et al.*, 1997a).

10.5. Interim Verification of Improvement of Slurry

10.5.1. *Verification from Analysis of Settlement Data*

During a period of about a year, significant settlement was recorded. Based on the measured settlement and the predicted ultimate settlement obtained from both the hyperbolic methods (Sridharan and Sreepada, 1981) and Asaoka method (Asaoka, 1978), the degree of consolidation for the pilot test area was estimated to be more than 80%. These observational methods grossly overestimated the true extent of the soil improvement.

10.5.2. *Verification from Analysis of Piezometer Monitoring Data*

Both pneumatic and electrical piezometers were monitored throughout the period. It was noted that excess pore pressure suddenly increased beyond the additional load in the initial stage of filling and no excess pore water pressure dissipation was recorded up to a period of 11 months for the top two piezometers and up to 2 weeks for the bottom-most piezometers. Large settlement occurred during this period without pore pressure dissipation. The variations of excess pore water pressure and settlement with time are shown in Figs. 10.27–10.29. The interpretation of the excess pore pressure took into consideration the change in elevation of the piezometer tips obtained from the settlement of the extended pipe of the piezometer guard shell. A small decrease in the piezometric head was registered which was probably due to the reduction of load as a result of the submergence of the sand fill. This is shown in Fig. 10.30. To confirm this phenomenon, equilibrium pore pressure measurements were monitored with CPT long-term holding tests (Bo *et al.*, 1997a). The holding test was maintained until pore pressure readings reached equilibrium at the time of measurement. The pore pressures recorded from the CPT holding tests were found to agree with those measured by the piezometers, thus confirming that the piezometer measurements were correct. The measurements from piezometers and CPT holding tests are shown in Fig. 10.31. The degree of consolidation and effective stress at different elevations were calculated and compared with those computed from the settlement monitoring data. The

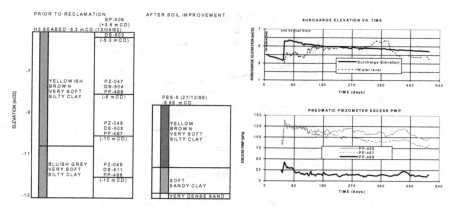

Fig. 10.27. Excess pore pressure versus time at silt pond pilot area (Pneumatic Piezometers).

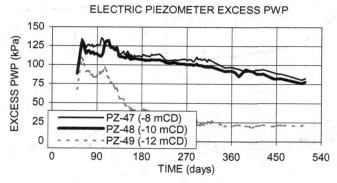

Fig. 10.28. Excess pore pressure versus time at silt pond pilot area (Vibratory Wire Piezometer).

results obtained were found to be far lower than that determined from the settlement monitoring data.

10.5.3. *Verification Using In Situ Tests*

Field vane shear tests and CPTs were carried out 14 months after surcharge to verify the true extent of soil improvement. Undisturbed samples were also collected from boring. The shear strength, overconsolidation ratio (OCR), and effective stress were then estimated from the results of *in situ* and laboratory tests. Details on the method of estimating these parameters

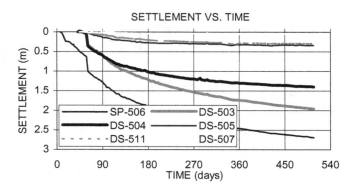

Fig. 10.29. Settlement versus time at silt pond pilot area.

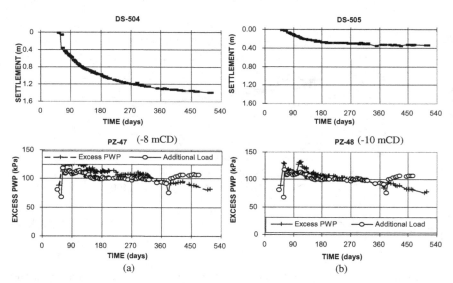

Fig. 10.30. Reduction of load caused by submergence of land compared with excess pore pressure from vibrating wire piezometer (a) No. 47 (b) No. 48.

from *in situ* tests have been discussed by Bo *et al.* (1997b). The degree of consolidation from these methods were compared with those obtained from the settlement measurements.

It was found that all parameters such as shear strength, OCR, and effective stress interpreted from *in situ* test data and laboratory tests were much lower than the values calculated from settlement analysis using the

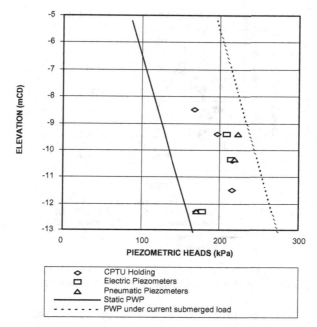

Fig. 10.31. Piezometric head measured from piezometers and CPT holding tests at 11 months after surcharge.

hyperbolic and Asaoka methods but were in reasonable agreement with the values obtained from the measured pore pressures.

Moreover, it confirmed that the soil had only consolidated to the low degree of consolidation indicated by the piezometers. Comparison of OCR, shear strength, effective stress, and degree of consolidation are shown in Figs. 10.32, 10.33, 10.34, and 10.35, respectively.

Therefore, it can be concluded that the large settlement that had occurred did not lead to any significant increase in effective stress or strength.

10.6. Verification of the Proposed Large Strain Deformation Model

The large strains occurred during the early stage of deformation of the silt pond slurry with little or no pore pressure dissipation cannot be explained by Terzaghi's conventional theory of consolidation. The top 5.3 m of material was in a slurry state and the bottom 2.5 m had already become

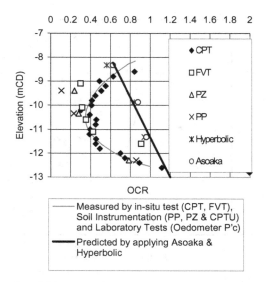

Fig. 10.32. OCR versus elevation interpreted from various tests.

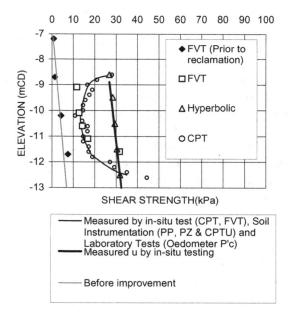

Fig. 10.33. Comparison of shear strength measured from various tests.

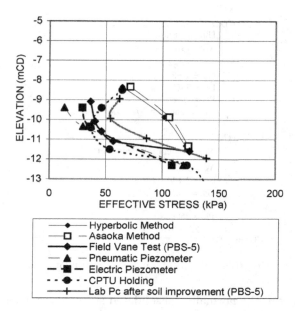

Fig. 10.34. Comparison of effective stress measured from various tests.

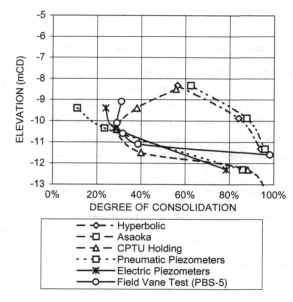

Fig. 10.35. Comparison of degree of consolidation interpreted from various tests.

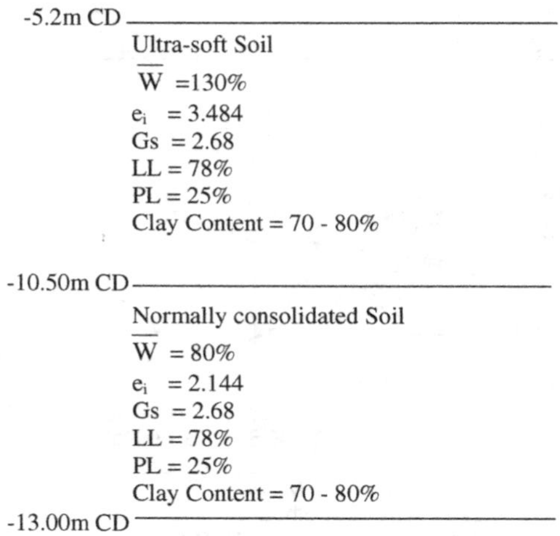

```
                  -5.2m CD ─────────────────────────────
                           Ultra-soft Soil
                           W̄ =130%
                           eᵢ = 3.484
                           Gs = 2.68
                           LL = 78%
                           PL = 25%
                           Clay Content = 70 - 80%

                 -10.50m CD ─────────────────────────────
                           Normally consolidated Soil
                           W̄ = 80%
                           eᵢ = 2.144
                           Gs = 2.68
                           LL = 78%
                           PL = 25%
                           Clay Content = 70 - 80%
                 -13.00m CD ─────────────────────────────
```

Fig. 10.36. Profile of soil at silt pond pilot area.

a Terzaghi soil in the normally consolidated state. The settlement of both layers were studied and analyzed. Figure 10.36 shows the profile of soil together with the geotechnical parameters of the silt pond pilot area.

10.6.1. *Determination of Compression Indices*

The compression indices at the different log cycles of pressure were obtained from Eqs. (7.11) to (7.15) using the liquid limit. The final void ratios for the relevant additional loads were calculated using Eqs. (7.18) and (7.19). The compression indices, void ratios at the liquid limit, and void ratio at 10 kPa for the ultra-soft upper layer are summarized in Table 10.2. The compression indices of normally consolidated lower soil layer are also

Table 10.2. Summary of compressibility parameters for ultra-soft upper and lower soil layer at silt pond area.

Parameters	Ultra-soft upper layer	Lower soil layer
Void ratio at liquid limit (e_L)	2.09	—
Void ratio at 10 kPa (e_{10})	2.20	—
C_{c1}^*	1.28	—
C_{c2}^*	0.19	0.91
C_{c3}^*	0.49	0.49

Table 10.3. Summary of predicted settlement under various stages at silt pond pilot area.

Stages	Addt'l. load (kPa)	Load duration (t) (month)	Ultimate settlement for upper layer in (m)	Settlement at time t for upper layer in (m)	Ultimate settlement for lower layer in (m)	Settl't at time t for lower layer in (m)	Predicted total settlement at time t (m)	Measured total settlement at time t (m)
1	65.20	2.00	2.174	0.472	0.630	0.130	0.602	0.25
2	105.45	28.70	2.416	2.408	0.720	0.717	3.125	3.10
3	186.45	16.53	2.585	2.584	0.806	0.805	3.389	3.40

included in Tables 10.2. and 10.3 shows the final expected load at various construction stages and the predicted settlement at the end of each stage together with the total cumulative settlement. For the ultra-soft soil, the initial effective stress is taken as 1 kPa and for normally consolidated soil the initial stress is calculated using $\sigma' = \gamma h$.

Submergence of the fill due to settlement was taken into consideration in the estimation of the final expected pressure. The void ratio and thickness of the compressible layer were updated at each stage of loading. The calculated final settlement was in close agreement compared with the measured settlement. The time rate of settlement was predicted using the proposed finite difference model. Construction stages and duration of load in various stages were modeled as shown in Fig. 10.37. The pore pressure versus time for the various cases are shown in Figs. 10.38 and 10.39.

Various C_F values were applied to the different loading stages. The applied C_F values for each stage are shown in Table 10.4.

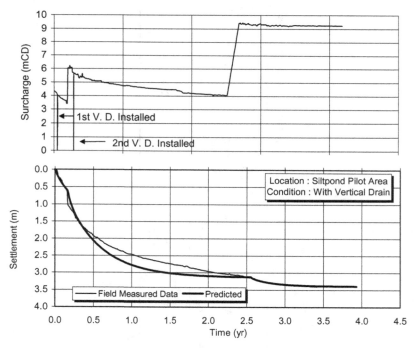

Fig. 10.37. Construction sequence for pilot area and verification using data from silt pond pilot area.

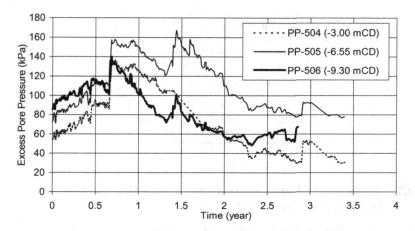

Fig. 10.38. Excess pore pressure versus time at large settlement area.

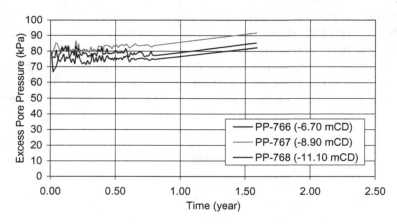

Fig. 10.39. Excess pore pressure versus time for no-drain area.

Several passes of vertical drain at 2 m square spacing were installed. This is because after the vertical drains have performed for a certain duration, the large strains and buckling of the drains would have made them virtually ineffective by the time the next round of drains was installed. The time rates of settlement were calculated for the two separate layers and combined to give the total settlement. It can be seen in Fig. 10.37 that the time rates of settlement predicted using the proposed model closely match the measured data.

Table 10.4. Summary of applied C_F values in the prediction of time rate of settlement in silt pond pilot area.

Stages	C_F (m²/yr)	
	Upper layer	Lower layer
1	0.0287	0.0539
2	0.0903	0.1264
3	0.1733	0.1827

10.7. Verification Using Data from Area with the Largest Deformation in the Main Works

Since verification of the reliability of the proposed equations and model using one set of test data may not be sufficiently conclusive, prediction was made for another area in the main works where the largest settlement occurred (see Fig. 10.40). The physical parameters and compression indices of the study area are shown in Table 10.5. Figure 10.41 shows the soil profile and geotechnical parameters of the area.

Again, the compression indices for the various pressure ranges were obtained from the proposed equations using the measured liquid limits. The computed settlements were found to be in close agreement with the

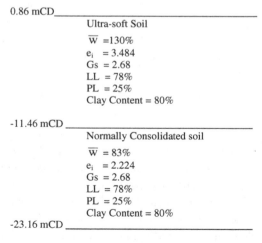

Fig. 10.40. Profile of soil at the area with the largest settlement in the silt pond.

Table 10.5. Summary of compressibility parameters for ultra-soft upper and lower soil layer at silt pond large settlement area.

Parameters	Ultra-soft soil	Lower soil layer
Void ratio at liquid limit (e_L)	2.09	—
Void ratio at 10 kPa (e_{10})	2.20	—
C_{c1}^*	1.28	—
C_{c2}^*	0.912	0.92
C_{c3}^*	0.50	0.50

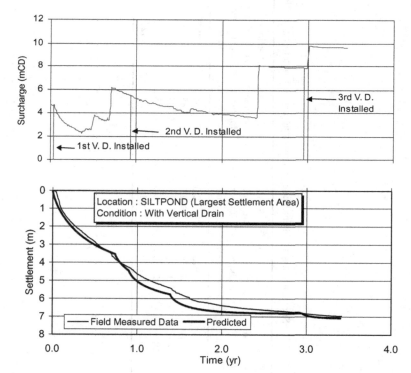

Fig. 10.41. Construction sequence for silt pond and verification using the data from the largest settlement area in the main work.

measured settlement (Table 10.6). The construction sequence is shown in Fig. 10.41.

The final expected loads and settlements at various stages are shown in Table 10.6. The measured and predicted time rates of settlement are in

Table 10.6. Summary of settlement under various stages at silt pond large settlement area.

Stages	Additional load (kPa)	Load duration (t) (months)	Ultimate settlement for upper layer in (m)	Settlement at time (t) for upper layer in (m)	Ultimate settlement for lower layer in (m)	Settlement at time t for lower layer in (m)	Predicted total settlement at time t (m)	Measured total settlement at time t in (m)
1	52.8	9.0	3.579	2.030	1.491	1.491	3.521	3.40
2	72.8	2.0	3.974	2.700	1.748	1.748	4.448	4.30
3	93.2	6.0	4.277	3.697	2.068	2.072	5.767	5.40
4	128.6	19.0	4.533	4.530	2.278	2.278	6.810	6.70
5	159.6	5.5	4.678	4.651	2.422	2.424	7.075	6.95

Table 10.7. Summary of applied C_F values used in the prediction of time rates of settlement in silt pond large settlement area.

Stages	C_F (m²/yr)	
	Upper layer	Lower layer
1	0.0465	0.2758
2	0.1018	0.3605
3	0.0957	0.5492
4	0.1488	0.7155
5	0.1943	0.7804

close agreement, as shown in Fig. 10.41. The slight variation of time rate of settlement in each stage may be due to a slight variation of the applied C_F values at each stage with the field C_F values. The applied C_F values in each stage are shown in Table 10.7. It could be seen that the C_F values of the lower layer clay for this area are very different from those in the pilot area. In reality, C_F values vary with the applied total and effective stress at each step. It should be noted that C_F values of lower clay layer in this area for various stages and pilot area are much different due to the differences in the stress level. The slight variation in settlement magnitude at each stage could also be due to the variation in the estimated additional load at each stage.

10.8. Verification of Large Strain Deformation using Data from No-Drain Area

The verifications in the earlier sections were all included with vertical drains, therefore an additional verification was made for an area with no vertical drains. In addition, the area was constructed with a single stage loading to +4 m CD. Figure 10.42 shows the soil profile and geotechnical

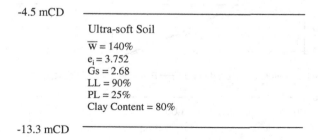

Fig. 10.42. Profile of soil at the no-drain area.

parameters of the no-drain area. In this area the filling was carried out on the freshly filled ultra-soft soil.

The parameters of the slurry in the no-drain area are shown in Table 10.8. The applied large strain coefficient of consolidation and

Table 10.8. Summary of compression parameters for ultra-soft soil in the containment bund, no-drain area.

Parameter	Unit	Value
Natural water content	%	140
Specific gravity	—	2.68
Initial void ratio	—	3.75
Liquid limit	%	90
Plastic limit	%	25
Clay content	%	80
C_{c1}^*	—	1.27
C_{c2}^*	—	1.03
C_{c3}^*	—	0.58

Table 10.9. Summary of applied C_F values and predicted settlement at the no-drain area.

Additional load (kPa)	Load duration t in (months)	Applied C_F (m²/yr)	Ultimate settlement in (m)	Settlement at time t in (m)	Measured settlement at time t in (m)
88	24	0.0425	4.15	1.042	1.055

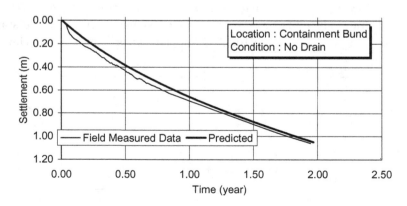

Fig. 10.43. Verification using data from the no-drain area in the main work.

predicted settlement using the proposed equations are shown in Table 10.9 together with the measured data. The settlement at time t can be calculated using the void ratio change data. The time rate of settlement predicted from the proposed model and measured data are compared in Fig. 10.43. It can be seen that the curves are in good agreement.

10.9. Summary

Although the ultra-soft soil had almost no strength, reclamation of the silt pond was successfully carried out with a sand spreading method. The sand fill was successfully placed with minimum failure due to the slow process of sand spreading which enabled the additional load to be imposed gradually and evenly. Geofabric helped in strengthening the ultra-soft soil foundation to a certain extent. Due to the occurrence of large strain, buckling of the vertical drains prevailed which caused the performance of the vertical drains to deteriorate. Therefore, installation of two or more passes of vertical drains was necessary to facilitate the dissipation of pore pressure.

Based on the monitoring data from three sets of soil instrument clusters in the slurry pond, it was found that large deformations would occur with little or no gain in effective stress and strength in the ultra-soft slurry-like soil upon the application of additional load from the sand fill. This was confirmed by *in situ* tests as well as laboratory tests. The magnitude of large strain consolidation in the ultra-soft soil can be predicted using the proposed equation which only requires the knowledge of compression indices at various stages. The required parameters, C_{c1}^*, C_{c2}^*, and C_{c3}^*, can be obtained from the proposed relationship with void ratio at the liquid limit. The predicted magnitude of settlement agreed very well with the measured data.

The time rate of settlement can be reasonably predicted using the proposed finite difference solution. The predicted time rate of settlement was in good agreement with the measured data. In the proposed finite difference program, the initial variation of void ratio along the profile as well as changes in void ratio and the large strain coefficient of consolidation in staged construction and various loading ranges can be taken into consideration. The computer program can be used in situation with and without drains. For the case with vertical drain, the presence of vertical drain is simulated by specifying proper vertical drainage boundaries within the soil mass.

Chapter 11

SUMMARY

The formation of soil was reviewed in this book, starting from sedimentation to completion of self-weight consolidation. The one-dimensional deformation of ultra-soft soil which is still undergoing self-weight consolidation upon loading was investigated with various types of laboratory tests using different equipments. Equations for obtaining various proposed compression indices and predicting the magnitude of settlement were validated with laboratory measurements. A finite difference model applying the large strain theory by Gibson *et al.* (1967) was also validated with laboratory and field measurements.

11.1. Sedimentation and Consolidation

The formation of fine-grained soil typically goes through sedimentation and self-weight consolidation. Although the process commences with sedimentation, these two processes occur simultaneously. The self-weight consolidation starts from the bottom while the sedimentation process continues at the top. There are four major transition points during the soil-forming process. The first one is where the effective stress becomes measurable. The second and third points are the void ratios at the end of sedimentation and self-weight consolidation, respectively. The last point corresponds to a state where the very soft slurry-like soil becomes a Terzaghi soil when effective stress controls the deformation process. For an ultra-soft soil which is still undergoing self-weight consolidation, the deformation behavior due to an applied load is different from that of loading on a natural soil. Therefore a trilinear effective stress model is proposed to predict the magnitude of settlement of an ultra-soft soil under an applied load.

11.2. Experimental Studies

A large diameter consolidometer was used to simulate the field condition. Pore pressure at various locations and settlement at the top were measured. Test results indicated that there was little or no pore pressure dissipation during the early stages of deformation. The average transition void ratio was determined, based on the commencement of pore pressure dissipation at various locations in the sample. It was however difficult to obtain an e-log σ'_v relationship and compression parameter from this type of test because of the large variation of effective stress within the soil sample.

One-step instantaneous loading tests were carried out with a small-scale consolidometer on samples with different initial moisture contents. It was observed that settlement took place without any appreciable dissipation of pore pressure. The approximate transition point when the slurry turned into a Terzaghi soil can be determined by various methods based on either rate of settlement, strain, and hydraulic conductivity reduction or the convergence of void ratio change-vs-time curves. A slightly more accurate transition point can be determined based on the pore pressure measurements at the base of the sample. The average transition void ratio determined based on the settlement analysis was lower than that determined from pore pressure consideration. An accurate transition point can be determined only if settlement and pore pressure at a particular soil element can be measured accurately.

In the laboratory tests with the hydraulic consolidation cell under instantaneous loading, even thin ultra-soft soil samples of less than 40 mm registered no significant pore pressure dissipation for a period of around 200 min. The transition points could be determined based on settlement and pore pressure data. With thinner samples, the transition points determined from settlement and pore pressure were in closer agreement. However, the void ratio at the transition point determined from pore pressure was slightly higher than that determined from settlement since the pore pressure was measured at the base of the sample.

From the e-log σ'_v curves obtained for the step loading tests with the hydraulic Rowe Cell, three different compression indices (C^*_{c1}, C^*_{c2}, and C^*_{c3}) could be determined from the three log cycles of stresses between 1 and 1000 kPa. For soils with similar liquid limit and mineralogy, similar C^*_{c3} were obtained regardless of the initial moisture content. The value was the same as the intrinsic compression index C^*_c proposed by Burland (1990). However, C^*_{c1} was highly dependent on the initial moisture content.

Therefore, a family of e-log σ'_v curves in the first log cycle would be obtained from the same type of soil with different initial moisture contents.

The void ratio at 10 kPa (e_{10}) which is an important parameter for determining various compression indices could be estimated from the void ratio at the liquid limit. Equations for settlement estimation in the slurry and soil stages were proposed using the three compression indices and the e_{10} value. The computed and measured void ratios and settlements under various load steps were validated with laboratory measurements and found to be in very good agreement.

Alternatively, the e-log σ'_v relationship can be obtained from the Constant Rate of Loading (CRL) tests when an appropriate rate of loading is used. The e-log σ'_v curves can be used to obtain compression parameters. Hydraulic conductivity (k) and large strain coefficient of consolidation (C_F) can also be interpreted from the CRL test. A loading rate of 1 kPa/800 s was found to be suitable for obtaining a reasonable e-log σ'_v curve from the CRL test on ultra-soft soil.

Another method of testing to obtain the compression parameters, k and C_F for ultra-soft soil is the Constant Rate of Strain (CRS) test. Only tests with slow strain rate of 0.01%/min could produce more or less constant excess pore pressure ratio below 30% and yield a reasonable e-log σ'_v curve. An α value of 0.75 for calculating the average effective stress from pore pressures measured at the base is adopted for the CRL and CRS tests.

It is recommended that more step-loading tests with hydraulic Rowe Cell, CRL or CRS can be conducted on various ultra-soft soils with different mineralogy and liquid limits to study the compression behavior of different types of ultra-soft soils. Small incremental loads, low strain, and loading rates should be carried out with pore pressure and displacement measurements using sensitive and accurate transducers.

11.3. Validation of Proposed Finite Difference Model

A numerical model based on the implicit finite difference method was proposed. This model uses the governing equation of large strain theory derived by Gibson et al. (1967) and the newly proposed equation using various compression indices. Results from the proposed finite difference model were validated against laboratory measurements from step-loading tests with hydraulic Rowe Cell. The computed magnitude and time rate of settlements were in good agreement with the measured values at all stress levels in both slurry and soil stages. For settlement prediction, the proposed model requires only the compression indices in the three log cycles which could be easily obtained from compression tests or from the relationship with void ratio at the liquid limit.

Comparisons were made between the measured time rate of settlement and those computed by the proposed methods as well as by methods proposed by Cargill (1982) and Lee and Sills (1981). Different C_F values were used depending on the magnitude of stress at various levels. It was found that although the predictions from Cargill (1982) generally agreed with the measured data, this method required the selection of a relevant N value in order to obtain a correct time factor curve. Agreement was not very good at low stress levels since N values were not constant and varied with stress levels. The method by Lee and Sills (1981) is simpler. It has only one time factor curve for each type of drainage condition. The predicted and the measured values were in good agreement with those at low stress levels. The present proposed model yielded the best agreement with the measured magnitude and rate of settlement including those at low stress levels. The time factor curves from the proposed model were similar to those proposed by Lee and Sills (1981) and Cargill (1982) with $N = 2$.

11.4. Case Study

The field performance of the reclamation and soil improvement of a silt pond involving very soft slurry-like soils was studied. No pore pressure dissipation was registered in the initial stage up to 11 months after fill placement and installation of vertical drains, even though 3 m of settlement was recorded.

Although the large measured settlement suggested a high degree of consolidation, the pore pressure measurements, however, indicated little pore pressure dissipation. The equilibrium pore pressure measurements using CPT long-term holding tests also confirmed that the degree of consolidation was indeed very small. Field vane tests, CPTs, and laboratory tests confirmed that there was virtually no gain in effective stress and undrained shear strength during the first 11 months after loading. It was therefore concluded that the deformation of ultra-soft soil did not follow Terzaghi's conventional consolidation theory.

The rate and magnitude of settlement for the drain and no-drain areas at the silt pond were predicted using the proposed model. The compression indices and large strain coefficient of consolidation parameters were estimated from the proposed correlations described in Chapters 7 and 8. It was found that the predicted magnitude and rate of settlement were in reasonably good agreement with the measured values at various stages of filling in both the areas with and without vertical drains. The deformation of ultra-soft soil under an applied load does not follow Terzaghi's effective stress theory. There was little pore pressure dissipation with high magnitude of settlement in the early stage of deformation. The newly proposed compression parameters and equations can be used to estimate the settlement magnitude. These parameters could be obtained from CRL, CRS, or Rowe Cell consolidation tests. The time rate of settlement could also be predicted with the finite difference program developed in this study.

REFERENCES

Aboshi, H., Yoshikuni, H. and Maruyama, S., Constant loading rate consolidation test, *Soils Found.*, **X**(1) (1970) pp. 43–56.

Aboshi, H., An experimental investigation on the sinilitude in the consolidation of a soft clay, including secondary creep settlement, *Proc. 8th ICSMF (Moscow)*, **4**(3) (1973) p. 88.

Abu-Hejleh, A. N., Znidarcic, D. and Barnes, B., Consolidation characteristics of phosphatic clays, *J. Geotech. Eng.*, **122**(4) (1996) pp. 295–301.

Anwar, E. Z., Wissa, A. M., Christian J. T., Davis, E. H. and Sigurd H., Consolidation at constant rate of strain, *J. Soil Mech. Found. Div.*, (1971) pp. 1393–1413.

Asaoka, A., Observational procedure of settlement prediction, *Soils and Found.*, **18**(4) (1978) pp. 87–101.

Barden, L. and Berry, P. L., Consolidation of normally consolidated clay, *Proc. Amer. Soc. Civil Eng.*, **SM 5**(4481) (1965) pp. 15–35.

Barron, R. A., Consolidation of fine grained soils by drain wells, *Trans. ASCE*, **113** (1948) pp. 718–734.

Been, K. and Sills, G. C., Self-weight consolidation of soft soils, An Experimental and Theoretical Study, *Géotechnique*, **31**(4) (1981) pp. 519–535.

Biot, M. A., Theory of elasticity and consolidation for a porous anisotropic solid, *J. Appl. Phys.*, **20**(2) (1955) pp. 182–185.

Bo M. W., Arulrajah, A. and Choa, V., Large deformation of slurry-like soil due to additional load, in *Int. Con. Found. Failure*, May (1997a), Singapore, pp. 289–296.

Bo M. W., Arulrajah, A. and Choa, V., Assessment of degree of consolidation in soil improvement project, in *Int. Conf. Ground Improvement Tech.*, May (1997b), Macau, pp. 71–80.

Bo M. W., Arulrajah, A. and Choa, V., Large deformation of slurry-like soil, in *Deformation and Progressive Failure in Geomechanics*, eds. Asaoka, A., Adachi, T. and Oka (Balkema, 1997c), pp. 437–442.

Bo M. W., Arulrajah, A., Choa, V. and Na, Y. M., Land reclamation on slurry-like soil foundation, in *Problematic Soils*, eds. Yonagisawa, Morota and Mitachi (Balkema, 1998), pp. 763–766.

Bo M. W., Arulrajah, A., Choa, V. and Na, Y. M., One-dimensional compression of slurry with radial drainage, *Soils and Found.*, **39** (1999).

Bo M. W., Choa, V. and Wong, K. S., Compression test on slurry with small scale consolidometer, *Can. Geotech. J.*, (2002a).

Bo M. W., Choa, V., Wong K. S. and Teh, C. I., Investigation on deformation behaviour of high moisture content soil, *Soils Found.*, **42**(2) (2002b).

Bowden, R. K., Compression behaviour and shear strength characteristics of a natural silty clay sedimented in laboratory, *D. Phil thesis*, Oxford University (1988).

Burghignoli, A., An experimental study of the structural viscosity of soft clays by means of continuous consolidation tests, in *7th European Conf. Soil Mech. Foundation Eng.*, Brighton, (1979) Vol. 2, pp. 23–28.

Bustos, M. C., A modification of the Kynch theory of sedimentation, *AIChE J.*, **33**(2) (1987) pp. 312–319.

Burland, J. B., On the compressibility and shear strength of natural clays, *Géotechnique*, **40**(3) (1990) pp. 329–378.

Cargill, K. W. and Schiffman, R. L., FSCONS-I, Version-I, Level-A, One dimensional finite strain consolidation, dimensionless void ratio for a thick, normally consolidated homogeneous layer, *Geotechnical Engineering* (Software Activity Department of Civil Engineering, University of Colorado, Boulder, 1980).

Cargill, K. W., Consolidation of soft layers by finite strain analysis, Miscellaneous Paper GL-82-3 (*US Army Engineer Waterways Experiment Station*, Vicksburg, Miss, 1982a).

Cargill, K. W., Consolidation behaviour of fine-grained dredged material, *Technical Report D-83-1* (*US Army Engineer Waterways Experimental Station*, Vicksburg, Miss, 1982b).

Cargill, K., Prediction of consolidation of very soft soil, *J. Geotech. Eng.*, **110**(6) (1984) pp. 775–795.

Carillo, N., Simple two and three dimensional cases in the theory of consolidation of soils, *J. Math. Phys.*, **21** (1942) pp. 1–5.

Carrier, W. D. III and Keshian, B. Jr., Measurement and prediction of consolidation of dredged material, *12th Annual Dredging Seminar*, Texas A&M University.

Carrier, W. D. III and Bromwell, L. G., Geotechnical analysis of confined soil disposal, in *Proc. Ninth World Dredging Conf.*, Vancouver, Canada, (1980), pp. 313–324.

Carrier, W. D. III, Bromwell, L. G. and Somogyi, F., Design capacity of slurried mineral waste, in *Proc. Amer. Soc. Civil Engrs., J. Geotech. Eng. Div.*, **109**(5) (1983) pp. 699–716.

Carrier, W. D. III and Bromwell, L. G., Disposal and reclamation of mining and dredging wastes, in *Proc. Seventh Panamer. Conf. Soil Mech. Found. Eng.*, Vancouver, (1983), Vol. 2, pp. 727–738.

Carrier, W. D. III and Beckman, J. F., Correlations between index tests and the properties of remoulded clays, *Géotechnique*, **34**(2) (1984) pp. 211–228.

Casagrande, A., The determination of the pre-consolidation load and its practical significance, in *Proc. First Int. Conf. Soil Mech.*, Cambridge, Mass, (1936), Vol. 3, pp. 60–64.

Coe, H. S. and Clevenger, G. H., Methods for determining the capacities of sline-settling tanks, *Trans. Amer. Institute Mining Eng.*, **55** (1916) pp. 356–384.

Crawford, C. B., The resistance of soil structure to consolidation, *Can. Geotech. J.*, **2**(2) (1965) pp. 90–97.

Cryer, C. W., A comparison of the three-dimensional consolidation theories of Biot and Terzaghi, *Quart. J. Mech. Appl. Math.*, **16** (1963) pp. 401–412.

de Josselin de Jong, G., Consolidation models consisting of an assembly of viscous elements or a cavity channel network, *Géotechnique*, **81**(2) (1968) p. 195.

Dixon, D. C., Momentum-balance aspects of free settling theory I: Batch thickening, *Separation Sci.*, **12** (1977a) pp. 171–192.

Dixon, D. C., Momentum-balance aspects of free settling theory II: Continuous steady-state thickening, *Separation Sci.*, **12** (1977b) pp. 193–200.

Fitch, B., Sedimentation process fundamentals, *Trans. Amer. Inst. Mining Eng.*, **223** (1962) p. 129.

Fitch, B., Current theory and thickener design, *Ind. Eng. Chem.*, **58** (1966a) p. 18.

Fitch, B., A mechanism of sedimentation, *Ind. Eng. Chem. Fundament.*, **5**(1) (1966b) pp. 129–131.

Fitch, B., Unresolved problems in thickener design and theory, 29th in *Res. Conf. Filtrat. Separat. Soci. Chem. Engrs.* Japan (1972).

Fitch, B., Current theory and thickness design, *Filtrat. Separat.*, **12** (1975) pp. 335, 480, 636.

Fitch, B., Sedimentation of flocculating suspensions — State of the art, *AIChE. J.*, **25** (1979) pp. 913–930.

Fitch, B., Kynch theory and compression zones, *AIChE J.*, **29**(6) (1983) pp. 940–947.

Garlanger, J. E., The consolidation of soils exhibiting creep constant, effective stress, *Géotechnique*, **22**(1) (1972) pp. 71–78.

Gibson, R. E., The progress of consolidation in a clay layer increasing in thickness with time, *Géotechnique*, **8** (1958) pp. 171–182.

Gibson, R. E. and Lo, K. Y., A theory of consolidation for soils exhibiting secondary compression, *Norwegian Geotech. Institute*, Publication 41.

Gibson, R. E., England, G. L. and Hussey, M. J. L., The theory of one-dimensional consolidation of saturated clays, I. Finite non-linear consolidation of thin homogeneous layers, *Géotechnique*, **17** (1967) pp. 261–273.

Gibson, R. E., Schiffman, R. L. and Cargill, K. W., The theory of one-dimensional consolidation of saturated clays, II. Finite non-linear consolidation of thick homogeneous layers, *Can. Geotech. J.*, **18** (1981) pp. 280–293.

Gorman, C. T., Hopkins, T. C., Deen, R. C. and Drnevich, V. P., Constant rate of strain and controlled-gradient consolidation testing, *Geotech. Test. J.*, **1**(1) (1978) pp. 3–15.

Hansbo, S., Jamiolkowski, M. and Kok, L., Consolidation by vertical drains, *Géotechnique*, **31**(1) (1981) pp. 45–66.

Hight, D. W., Jardine, R. J. and Gens, A., The behavior of soft clays, Embankment on Soft Clays, Public Works Research Center, *Athens*, (1987), Chapter 2, pp. 33–158.
Huerta. A., Kriegsmann, G. A. and Krizek, R. J., Permeability and compression of slurries from seepage induced consolidation, *J. Geotech. Eng.*, **114**(S) (1988) pp. 614–627.
Imai, G., Tsuruya, K. and Yano, K., A treatment of salinity in water content determination of very soft clay, *Soils Found.*, **19**(3) (1979) pp. 84–89.
Imai, G., Experimental studies on sedimentation mechanism and sediment formation of clay materials, *Soils Found.*, **21**(1) (1981) pp. 7–21.
Imai, G., Analytical examination of the foundations to formulate consolidation phenomena with inherent time-dependence, *Compression and Consolidation* (Balkema Rotterdam, 1995), pp. 891–935.
Irwin, M. J., Consolidation testing with a constant rate of loading, *Leaflet LF466* (Transport and Road Research Laboratory, Crowthorne, Berks, England, 1975).
Janbu, N., Tokheim, O. and Senneset, K., Consolidation tests with continuous loading, *Proc. 10th ICSMFE.*, Stockholm, (1981), Vol. 1, pp. 645–654.
Jardine, R. J., Potts, D. M., Fourie, A. B. and Burland, J. B., Studies of the influence of non-linear stress–strain characteristics in soil structure interaction, *Géotechnique*, **36**(3) (1986) pp. 377–396.
Jardine, R. J. and Hight, D. W., The behavior and analysis of embankments on soft clay, *Embankments on Soft Ground* (Public Works Research Center, Athens, 1987), Ch. 3, pp. 195–224.
Jardine, R. J., St. John, H. D., Hight, D. W. and Potts, D. M., Some practical applications of a non-linear ground model, in *Proc. 10th ECSMFE*, Florence, (1991), Vol. 1, pp. 223–228.
Katagiri, M. and Imai, G., A new in laboratory method to make homogenous clayey samples and their mechanical properties, *Soils Found.*, **34**(2) (1994) pp. 87–93.
Katagiri, M., Effect of clay water content on the consolidation of a sediment layer of reclaimed clay, in *Compression and Consolidation of Clayey Soil*, eds. Yoshikuni and Kusakabe (Balkema, Rotterdam, 1995), pp. 261–266.
Kearsey, H. A. and Gills, L. E., A study of sedimentation of flocculated theory slurries using a gamma-ray technique, *Trans. Institution Chemical Engrs.*, (1963) pp. 296–306.
Kim, H., Hirokane, S., Yoshikuni, H., Mosiwaki, T. and Kusakabe, O., Consolidation behaviour of dredged clay ground improved by horizontal drain method, in *Compression and Consolidation of Clayey Soils*, eds. Yoshikuni and Kusakabe (Balkema, Rotterdam, 1995), pp. 99–104.
Kynch, G. J., A theory of sedimentation, *Trans. Faraday Society*, **48** (1952) pp. 166–176.
Ladd, C. C., Foott, R., Ishihara, K., Schlosser, F. and Poulos, H. G., Stress deformation and strength characteristics, State of the Art Report, in *Proc. Ninth Int. Conf. Soil Mech. Foundation Eng.*, Tokyo (1977), Vol. 2, pp. 421–494.

Lacerda, W. A., Almcida, M. S. S, Santa Maria. P. E. L. and Coutinho, R. Q., Interpretation of radial consolidation tests, in *Compression and Consolidation of Clayey Soils* (Balkema, Rotterdam, 1995), pp. 1091–1096.
Lambe, T. W. and Whitman, R. V., *Soil Mechanics* (Wiley, New York, 1969).
Lee, K. and Sills, G. C., The consolidation of a soil stratum, including self-weight effects and large strains, *Int. J. Numer. Anal. Meth. Geomech.*, **5** (1981) pp. 405–428.
Leroueil, S., Samson, L. and Bozozuk, M., Laboratory and field determination of preconsolidation pressures at gloucester, *Can. Geotech. J.*, **20**(3) (1983) pp. 477–490.
Leroueil, S., Kabbaj, M., Tavenas, F. and Bouchard, R., Stress–strain–strain rate relation for the compressibility of sensitive natural clay, *Géotechnique*, **35**(2) (1985) pp. 159–180.
Leroueil, S., Kabbaj, M., Tavenas, F. and Bouchard, Stress–strain–strain rate relation for the compressibility of sensitive natural clays reply, *Géotechnique*, **36**(2) (1986a) pp. 288–290.
Leroueil, S., Recent development in consolidation of natural clays, *Canadian Geotechnical Journal*, **25**(1) (1988b) pp. 85–107.
Leroueil, S., Compression of clays: Fundamental and practical aspect, in *ASCE Conf. Vertical Horizontal Deformat. Found. Embankments*, College Station, (1994), Vol. 1, pp. 57–76.
Leroueil, S., Geotechnical characteristics of Eastern Canada clays, in *Characterization of Soft Marine Clays*, eds. Tsuchida and Nakase (Balkema, Rotterdam, 1999), pp. 3–32.
Leonards, G. A. and Ramiah, B. K., Time effects in the consolidation of clays, in *Symposium of Time Rates of Loading in Soil Testing* (ASTM, Special Technical Publication, 1960), Vol. 254, pp. 116–130.
Li, H. and Williams, D. J., Numerical modeling of combined sedimentation and self-weight consolidation of an accreting coal mine tailings slurry, *Proceedings, IS-Hiroshima*, (1995).
Liu, J. C. and Znidarcic, D., Modeling one dimensional compression characteristics of soils, *J. Geotech. Eng.*, **117**(1) (1991) pp. 162–169.
Mandel, J., Étude Mathématique de la consolidation des sols, in *Actes Du Colloque International De Mechanique*, Poitiers, France, Vol. 4, pp. 9–19.
Mandel, J., Consolidation des sols (Étude Mathématique), *Géotechnique*, **3** (1953) pp. 287–299.
Matsuda, H. and Aboshi, H., Discussion — A constitutive equation of one-dimension consolidation derived from inter-connected tests, *Soils Found.*, **33**(3) (1993) p. 153.
Matsuda, H. and Nagatani, T., Consolidation tests with constant rate of loading by separate-type consolidometer, in *Compression and Consolidation of Clayey Soils* (Balkema, Rotterdam, 1995), pp. 123–128.
McRoberts, E. C. and Nixon, J. F., A theory of soil sedimentation, *Can. Geotech. J.*, **13** (1976) pp. 294–310.
Mcvay M., Townsend, F. and Bloomquist, D., Quiescent consolidation of phosphatic waste clays, *Journal Geot. Engineering*, ASCE 112, pp. 1033–1049.

Mesri, G. and Rokshar, A., Theory of consolidation of clays, *Geotech. Eng. Div.*, ASCE, **100** (1974) pp. 889–904.
Mesri, G. and Choi, Y. K., The uniqueness of the end of primary (EOP) void ratio effective stress relationship, in *Proc. 11th IC SMFE*, (1985), Vol. 2, pp. 587–590.
Mesri, G., The fourth law of soil mechanics: The law of compressibility, in *Proc. Int. Symp. Geotech Eng. Soft Soils* (Sociedad Mexicana de Mecanica, de Suelos, Mendoza (c.d), 1987), Vol. 2, pp. 179–187.
Mesri, G., Shahien, M. and Feng, T. W., Compressibility parameters during primary consolidation, *Compression and Consolidation of Clayey Soils*, eds. Yoshikuni and Kusakabe (Balkema, Rotherdam, 1995), Vol. 2, pp. 1021–1037.
Michaels, A. S. and Bolger, J. C., Settling rates and sediment volumes of flocculated kaolin suspensions, *Indus. Engg. Chem. Fundamental I*, **24** (1962) pp. 24–33.
Mikasa, M., Consolidation analysis of soft clay considering self-weight of the clay and its application to the reclamation work, in *Proc. Annu. Convention JSCE*, 1961, pp. 7–9 (in Japanese).
Mikasa, M., Two basic questions on consolidation, *Compression and Consolidation of Clayey Soil*, eds. Yoshikuni and Kusakabe (Balkema, 1995), pp. 1097–1098.
Mikasa, M. and Takada, N., Non-linear consolidation theory for none homogeneous soil layers, in *Compression and Consolidation of Clayey Soil* (Balkema, 1995), Vol. 2, pp. 447–452.
Mischler, R. T., Settling slimes at the tigre mill, *Eng. Mining J.*, **94** (1912) p. 643.
Monte, J. L. and Krizek, R. J., One-dimensional mathematical model for large strain consolidation, *Géotechnique*, **26**(3) (1976) pp. 495–510.
Morgenstern, N. R. and Nixon, J. F., One-dimensional mathematical model for large-strain consolidation, *Géotechnique*, **26**(3) (1971) pp. 459–510.
Na, Y. M., Choa, V., Bo, M. W. and Arulrajah, A., Use of geosynthetics for reclamation on slurry-like soil foundation, in *Problematic Soils*, eds. Yanagisawa, Moroto and Mitachi (Balkema, Rotterdam, 1988), ISBN 9054109971, pp. 767–771.
Olson, R. E. and Mitronovas, F., Shear strength and consolidation characteristics of calcium and magnesium illite, *Clays Clay Miner*, **11** (1962) pp. 185–209.
Olson, R. E. and Mesri, G., Mechanism controlling the compressibility of clay, *J. Soil Mech. Found. Div.*, **96**(SM6) (1970) pp. 1863–1878.
Pane, V. and Schiffman R. L., A note on sedimentation and consolidation, *Géotechnique*, **35**(1) (1985) pp. 69–70.
Roberts, E. J., Thickening, art or science, *Mining Eng.* (1949) p. 161.
Rowe, P. W., A new consolidation cell, *Géotechnique*, **16**(2) (1966) p. 162.
Rowe, P. W., Embankments on soft alluvial ground, *Q. J. Engny Grol.*, **5** (1972) pp. 127–141.
Russel, W. B., Saville, D. A. and Schowalter, W. R., *Colloidal dispersions Cambridge* (Cambridge University Press, 1989).
Salem, A. M. and Krizek, R. J., Consolidation characteristics of dredging slurries, *J. Watways Harb. Coastal Eng. Div.*, **99** (1973) pp. 439–457.

Schiffman, R. L., Consolidation of soil under time dependent loading and varying permeability, in *Proc. Highway Research Board*, 1958, Vol. 37, pp. 584–615.

Schiffman, R. L., Chen, A. T. F. and Jordan, J. C., An analysis of consolidation theories, *J. Soil Mech. Found. Dir.*, **95**(SM1) (1969) pp. 285–312.

Schiffman, R. L., Finite and infinitesimal strain consolidation, *J. Geotech. Eng. Div.*, **106** (1980) pp. 203–207.

Schiffman, R. L. and Cargill, K. W., Finite strain consolidation of sedimenting clay deposits, in *Proc. 10th Int. Conf. Soil Mech.*, (1981) Vol. 1, pp. 239–242.

Schiffman, R. L., Pane, V. and Gibson, R. E., An overview of nonlinear finite strain sedimentation and consolidation, in *ASCE, Convention in Francisco*, California, (1984) pp. 1–29.

Scott, K. J., Mathematical models of mechanism of thickening, *I&E.C., Fundamentals*, **5** (1966) p. 109.

Scotts, J. D., Dusseault, M. B. and Carrier, D. W., Large scale self-weight consolidation testing, consolidation of soils, testing and evaluation, in *ASTM STP 892*, eds. Yong and Townsend (ASTM, Philadelphia, 1986), pp. 5000–5155.

Sheahan, T. C. and Watters, P. J. Using an automated rowe cell for constant rate of strain consolidation testing, *Geotech. Testing J.*, **19**(4) (1996) pp. 354–363.

Shirato, M., Kato, K., Kobayashi and Sakazaki, Analysis of settling thick slurries due to consolidation, *J. Chem. Eng.*, Japan, **3** (1970) p. 98.

Sills, G. C. and Been, K., Escape of pore fluid from consolidating sediment, in *Proc. Conf. Transfer Process. Cohesive Sediment Syst.* (Plenum Press, 1981).

Sills, G. C. Hoare, S. D. and Baker, N., An experimental assessment of the restricted flow consolidation tests, in *Proc. ASTM Symp. Consolidation Behaviour Soils*, October, (1985), Fort Lauderdale, Florida, USA.

Sills, G. C., Time dependent processes in soil consolidation, *Compression and Consolidation of Clayey Soils*, eds. Yoshikuni and Kusakabe (Balkema, Rotterdam, 1995), pp. 875–890.

Smith, R. E. and Wahls, H. E., Consolidation under constant rates of strain, *J. Soil Mech. Foundations Div. Proc. ASCE*, **95** (1969) pp. 519–539.

Sridharan, A. and Sreepada Rao, A., Rectangular hyperbolar fitting method for one-dimensional consolidation, *Geotech Testing J.* **C4**(4) (1981) pp. 161–168.

Talmage, W. P. and Fitch, E. B., Determining thickener unit areas, *I&E.C.* (1955) pp. 38, 47.

Tan, S. A., Tan, T. S., Ting, L. C., Yong, K. Y., Karunaratne, G. P. and Lee, S. L., Determination of consolidation properties for very soft clay, *Geotech. Test. J.*, **11**(4) (1988) pp. 233–240.

Tan, T. S., Yong, K. Y., Leong, E. C. and Lee, S. L., Sedimentation of clayey slurry, *J. Geotech. Eng.*, **116**(6) (1990a) pp. 885–898.

Tan, T. S., Yong, K. Y., Leong, E. C. and Lee, S. L., Behaviour of clay slurry, *Soils Found.*, **30**(4) (1990b) pp. 105–118.

Tan, T. S., Sedimentation to consolidation: A geotechnical perspective, in *Compression and Consolidation of Clayey Soil*, eds. Yoshikuni and Kusakabe (Balkema, Rotterdan, 1995), pp. 937–948.

Tanaka, Y., Non-uniformly consolidated ground around vertical drains, in *Proc. 30th year Anniversary Symp. Southeast Asian Geotech Soc.*, Bangkok, Thailand, (1997), pp. 1-232 to 1-243.

Tang, Y. X. and Imai, G., A constitutive relation with creep and its application to numerical analysis of one dimensional consolidation, in *Compression and Consolidation of Clayey Soils*, eds. Yoshikuni and Kusakabe (Balkema, Rotterdom, 1995), pp. 465–472.

Tavenas, F., Leblond, P., Jean, P. and Leroueil, S., The permeability of natural clay: Part I: Method of laboratory measurement, *Can. Geotech. J.*, **20** (1983a) pp. 629–644.

Tavenas, F., Jean, P., Leblond, P. and Leroueil, S., The permeability of natural clay, Part II: Method of laboratory measurement, *Can. Geotech. J.*, **20** (1983b) pp. 645–644.

Taylor, D. W. and Merchant, W., A theory of clay consolidation accounting for secondary compressions, *J. Math. Phys.*, **19** (1940) pp. 167–185.

Terzaghi, K., Die Berechnung der Durchlassigkeitsziffer des Tones aus dem verlang der Hydrodynomischen Spannungserscheinunpen, *Sitzungsbcrichte, Matematisch Notarwissenschaftliche Klasse* (Akademie der Wissenschaften, Vienna: 1923), Part 2.2, 132, 3/4, pp. 125–138.

Terzaghi, K., Erdbaumechanich auf Boden physikalische, *Groundlayer* (Leipzing, Deuticke, 1925).

Terzaghi, K. and Peck, R. B., *Soil Mechanics in Engineering Practice* (Wiley, New York, 1948).

Tiller, F. M., Revision of kynch Sedimentation theory, *AIChE J.*, **27** (1981) p. 823.

Toorman, E. A., Sedimentation and self-weight consolidation general unifying theory, *Géotechnique*, **46**(1) (1996) pp. 103–113.

Townsend, F. C. and McVay, M. C., SOA: Large strain consolidation predicting, *J. Geotech. Engrg.*, **116**(2) (1990) pp. 222–243.

Tsukada, Y. and Yasuhara, K., Scale effects in one-dimensional consolidation of clay, in *Compression and Consolidation of Clayey Soils*, eds. Yoshikuni and Kusakabe (Balkema and Rotterdam, 1995), Vol. 1, pp. 211–216.

Umehara, Y. and Zen, K., Constant rate of strain consolidation for very soft clayey soils, *Soils Found.*, **20**(2) (1980).

Van Essen, H. M., Greeuw, G. and Wichman, B., Combination of laboratory tests to determine consolidation parameter functions for sludge, in *Compression and Consolidation of Clayey Soils*, eds. Yoshikuni and Kusakabe (Balkema, Rotterdam, 1995), pp. 593–601.

Wichman, G. H. M., Consolidation behaviour of gassy mud: Theory and experimental validation, *Ph.D Thesis*, Technische Universiteit Delft (1999).

Wissa, A. E. Z., Christian, J. T., Davis, E. H. and Heiberg, S., Analysis of consolidation at constant strain rate, *J. Soil Mech. Found. Div. ASCE*, **97**, (SM10) (1971) pp. 1393–1413.

Woo, S. M., Brumrungsup, T. and Moh, Z. C., Effects of soil structure on compressibility of an artificially sedimented clay, *Proc. Int. Symp. Soft Clays*, Bangkok, Thailand, (1977), pp. 311–325.

Wood, D. M. and Wroth, C. P., The consolidation of some basic engineering properties of soil, in *Proc. Int. Conf. Behavior Offshore Struct.*, Trondheim, (1976), Vol. 2.

Wu, D., Improvement of clay slurry for land reclamation, *Ph.D.* (*Thesis*), Nanyang Technological University, Singapore (1994).

Yoshikuni, H., Kusakabe, O., Hirano T. and Ikegami, S., Elasto-viscous modeling of time-dependent behaviour of clay, in *Proc. the 13th Int. Conf.*, S.M.F.E. (1994), pp. 417–419.

Yoshikuni, H., Okada, M., Ikegami, S. and Hirano, T., One-dimensional consolidation analysis based on an elasto-viscous liquid model, in *Compression and Consolidation of Clayey Soils*, eds. Yushikuni and Kusakabe (Balkema, Rotterdam, 1995a), pp. 505–512.

Yoshikuni, H., Kusakabe, O., Okada, M. and Tajima S., Mechanism of one-dimensional consolidation, in *Proc. IS-Hiroshima* Vol. 1 (Compression and Consolidation of Clayey Soils) (Balkema, 1995b), pp. 497–504.

Znidarcic, D. and Schiffman, R. L., Finite strain consolidation test condition, Technical Note, *J. Geotech. Eng. Div.*, **107** (1981) pp. 684–688.